T0140141

Scalable Computing and Communications

Series editor

Albert Y. Zomaya
University of Sydney
New South Wales, Australia

More information about this series at http://www.springer.com/series/15044

Sourav Mazumder • Robin Singh Bhadoria
Ganesh Chandra Deka

Editors

Distributed Computing in Big Data Analytics

Concepts, Technologies and Applications

 Springer

Editors
Sourav Mazumder
IBM Analytics
San Ramon, CA, USA

Ganesh Chandra Deka
Directorate General of Training
Ministry of Skill Development and
   Entrepreneurship
New Delhi, Delhi, India

Robin Singh Bhadoria
Discipline of Computer Science and
   Engineering
Indian Institute of Technology Indore
Indore, Madhya Pradesh, India

ISSN 2520-8632   ISSN 2364-9496   (electronic)
Scalable Computing and Communications
ISBN 978-3-319-86713-7   ISBN 978-3-319-59834-5   (eBook)
DOI 10.1007/978-3-319-59834-5

Printed on acid-free paper

This Springer imprint is published by Springer Nature
The registered company is Springer International Publishing AG
The registered company address is: Gewerbestrasse 11, 6330 Cham, Switzerland

Editor's Notes

We are today living in the world of information explosion. Fortunately our human brain is reasonably fast and intelligent enough to capture relevant information from the large volume of data we are exposed to on every day basis. That helps us taking appropriate decisions and making right choices every moment in our business and personal lives. However, more and more, we have started facing difficulty in doing the same given that in many a case we need to take rapid decision after gathering insights from very high volume and numerous varieties of data. So having an aid in supporting human decision making process is becoming utterly important in today's world to make everyone's life easier and the decisions more accurate and effective. This aid is wh   at we otherwise call as Analytics.

The Analytics is anything but new to the human world. The earliest evidence of applying Analytics in business is found in late of seventeenth century. At that point of time Founder Edward Lloyd used the shipping news and information gathered from his coffee house to assist bankers, sailors, merchants, ship owners, and others in their business dealings, including insurance and underwriting. This made Society of Lloyds the world's leading market for specialty insurance for next two decades, as they could use historical data and proprietary knowledge effectively and quickly to identify risks. Next in early twentieth century human civilization saw few revolutionary ideas forming side by side in the area of Analytics both from academia as well as business. In academia, Moore's common sense proposition gave rise to the idea of 'Analytic Philosophy' which essentially advocates extending facts gathered from common place to greater insights. On the other hand, in the business side of the world, Frederick Winslow Taylor detailed out efficiency techniques in his book, The Principles of Scientific Management, in 1911, which were based on principles of Analytics. Also, during the similar time frame, the real life use of Analytics was actually implemented by Henry Ford by measuring pacing of the assembly line which eventually revolutionized the discipline of Manufacturing.

However, the Analytics started becoming more main mainstream, which we can refer as Analytics 1.0, with the advent of Computers. In 1944, Manhattan Project predicted behavior of nuclear chain reactions through computer simulations, in 1950 first weather forecast was generated by ENIAC computer, in 1956 shortest path problem was solved through computer based analytics which eventually transformed Air Travel and Logistics industry, in 1956 FICO created analytic model for credit risk prediction, in 1973 optimal price for stock options was derived using Black-Scholes model, in 1992 FICO deployed real time analytics to fight credit fraud and in 1998 we saw use of analytics for competitive edge in sports by the Oakland Athletics team. From the late 90's onwards, we started seeing major adoption of Web Technologies, Mobile Devices and reduction of cost of computing infrastructures. That started generating high volume of data, namely Big Data, which made the world thinking about how to handle this Big Data both from storage and consumption perspectives. Eventually this led to the next phase of evolution in Analytics, Analytics 2.0, in the decade of 2000. There we saw major resurgence in the belief in potential of data and its usage through the use of Big Data Technologies. These Big Data Technologies ensured that the data in any volume, variety and velocity (the rate at which it is produced and consumed) can be stored and consumed at reasonable cost and time. And now we are in the era of Big Data based Analytics, commonly called as Big Data Analytics or Analytics 3.0. Big Data Analytics is essentially about the use of Analytics in every aspect of human needs to answer the questions right in time, to help taking decisions in immediate need and also to make strategies using data generated rapidly in volume and variety through human interactions as well as by machines.

The key premise of Big Data Analytics is to make insights available to users, within actionable time, without bothering them of the ways the data is generated and the technology used to store and process the same. This is where the application of principles of Distributed Computing comes into play. The Distributed Computing brings two basic promises in the world of Big Data (and hence to Big Data Analytics) – ability to scale (with respect to processing and storage) with increase in volume of data and ability to use low cost hardware. These promises are highly profound in nature as they reduce the entry barrier for anyone and everyone to use Analytics and it also creates a conducive environment for evolution of analytics in a particular context with the change in business direction and growth.

Hence, to properly leverage benefits out of Big Data Analytics, one cannot undermine the importance of principles of Distributed Computing. The principals of Distributed Computing that involve data storage, data access, data transfer, visualization and predictive modeling using multiple low cost machines are the key considerations that make Big Data Analytics possible within stipulated cost and time practical for consumption by human and machines. However, the current literatures available in Big Data Analytics world do not cover the use of key aspects of Distributed Processing in Big Data Analytics in an adequate way which can highlight the relation between Big Data Analytics and Distributed Processing for ease of understanding and use by the practitioners. This book aims to cover that gap in the current space of books/literature available for Big Data Analytics.

The chapters in this book are selected to achieve the afore mentioned goal with coverage from three perspectives - the key concepts and patterns of Distributed Computing that are important and widely used in Big Data Analytics, the key technologies which support Distributed Processing in Big Data Analytics world, and finally popular Applications of Big Data Analytics highlighting how principles of Distributed Computing are used in those cases. Though all of the chapters of this book have the underlying common theme of Distributed Computing connecting them together, each of these chapters can stand as independent read so that the readers can decide to pick and choose depending on their individual needs.

This book will potentially benefit the readers in the following areas. The readers can use the understanding of the key concepts and patterns of Distributed Computing, applicable to Big Data Analytics while architecting, designing, developing and troubleshooting Big Data Analytics use cases. The knowledge of working principles and designs of popular Big Data Technologies in relation to the key concepts and patterns of Distributed Technologies will help them to select right technologies through understanding of inherent strength and drawback of those technologies with respect to specific use cases. The experiences shared around usage of Distributed Computing principles in popular applications of Big Data Analytics will help the readers understanding the usage aspects of Distributed Computing principals in real life Big Data Analytics applications-what works and what does not. Also, best Practices discussed across all the chapters of this book would be easy reference for the practitioners to apply the concepts in his/her particular use cases. Finally, in overall, all these will also help the readers to come out with their own innovative ideas and applications in this continuously evolving field of Big Data Analytics.

We sincerely hope that readers of today and future interested in Big Data Analytics space would find this book useful. That will make this effort worthwhile and rewarding. We wish all readers of this book the very best in their journey of Big Data Analytics.

Contents

On the Role of Distributed Computing in Big Data Analytics

Alba Amato

1 Introduction

Distributed paradigm emerged as an alternative to expensive supercomputers, in order to handle new and increasing users needs and application demands [1]. Opposed to supercomputers, distributed computing systems are networks of large number of attached nodes or entities connected through a fast local network [2]. This architectural design allows to obtain high computational capabilities by joining together a large number of compute units via a fast network and resource sharing among different users in a transparent way. Having multiple computers processing the same data means that a malfunction in one of the computers does not influence the entire computing process. This paradigm is also strongly motivated by the explosion of the amount of available data that make necessary the effective distributed computation. Gartner has defined big data as "high volume, velocity and/or variety information assets that demand cost-effective, innovative forms of information processing that enable enhanced insight, decision making, and process automation" [3]. In fact the huge size is not the only property of Big Data. Only if the information has the characteristics of either of Volume, Velocity and/or Variety we can refer the area of problem/solution domain as Big Data [4].Volume refers to the fact that we are dealing with ever-growing data expanding beyond terabytes into petabytes, and even exabytes (1 million terabytes). Variety refers to the fact that Big Data is characterized by data that often come from heterogeneous sources such as machines, sensors and unrefined ones, making the management much more complex. Finally, the third characteristic, that is velocity that, according to Gartner [5], "means both how fast data is being produced and how fast the data must be

A. Amato (✉)
Department of Industrial and Information Engineering, Second University of Naples,
Caserta, CE, Italy
e-mail: alba.amato@unina2.it; albaamato@gmail.com

© Springer International Publishing AG 2017
S. Mazumder et al. (eds.), *Distributed Computing in Big Data Analytics,*
Scalable Computing and Communications, DOI 10.1007/978-3-319-59834-5_1

1

processed to meet demand". In fact in a very short time the data can become obso-lete. Dealing effectively with Big Data "requires to perform analytics against the volume and variety of data while it is still in motion, not just after" [4]. IBM [6] proposes the inclusion of veracity as the fourth big data attribute to emphasize the importance of addressing and managing the uncertainty of some types of data. Striving for high data quality is an important big data requirement and challenge, but even the best data cleansing methods cannot remove the inherent unpredictabil-ity of some data, like the weather, the economy, or a customer's actual future buying decisions. The need to acknowledge and plan for uncertainty is a dimension of big data that has been introduced as executives seek to better understand the uncertain world around them [7]. Big Data are so complex and large that it is really difficult and sometime impossible, to process and analyze them using traditional approaches. In fact traditional relational database management systems (RDBMS) can not han-dle big data sets in a cost effective and timely manner. These technologies are typi-cally not enabled to extract, from large data set, rich information that can be exploited across of a broad range of topics such as market segmentation, user behavior profiling, trend prediction, events detection, etc. in various fields like pub-lic health, economic development and economic forecasting. Besides Big Data have a low information per byte, and, therefore, given the vast amount of data, the poten-tial for great insight is quite high only if it is possible to analyze the whole dataset [4]. The challenge is to find a way to transform raw data into valuable information.

So, to capture value from big data, it is necessary to use next generation innova-tive data management technologies and techniques that will help individuals and organizations to integrate, analyze, visualize different types of data at different spa-tial and temporal scales. Basically the idea is to use distributed storage and distrib-uted processing of very large data sets in order to address the four V's. There come the big data technologies which are mainly built on distributed paradigm. Big Data Technologies built using the principals of Distributed Computing, allow acquizition and analysis of intelligence from big data. Big Data Analytics can be viewed as a sub-process in the overall process of insight extraction from big data [8].

In this chapter, the first section introduces an overview of Big Data, describing their characteristics and their life cycle. In the second section the importance of Distributed Computing is explained focusing on issue and challenges of Distributed Computing in Big Data analytics. The third section presents an overview of tech-nologies for Big Data analytics based on Distributed Computing concepts. The focus will be on Hadoop.[1] which provides a distributed file system, YARN[2], a resource manager through which multiple applications can perform computations simultaneously on the data, and Spark,[3] an open-source framework for the analysis of data that can be run on Hadoop, its architecture and its mode of operation in comparison to MapReduce.[4] The choice of Hadoop is due to more elements. First

---

[1] hadoop.apache.org.

[2] https://hadoop.apache.org/docs/current/hadoop-yarn.html.

[3] spark.apache.org/.

[4] https://hadoop.apache.org/docs/r1.2.1/mapred_tutorial.html.

of all it is leading to phenomenal technical advancements. Moreover it is an open source project, widely adopted with an ever increasing documentation and community. In the end conclusion are discussed together with the current solutions and future trends and challenge.

2   History and Key Characteristics of Big Data

Distributed computing divides the big unmanageable problems around processing, storage and communication, into small manageable pieces and solves it efficiently in a coordinated manner [9]. Distributed computing are ever more widespread because of availability of powerful yet cheap microprocessors and continuing advances in communication technology. It is necessary especially when there are complex processes that are intrinsically distributed, with the need for growth and reliability.

Data management industry has been revolutionized by hardware and software breakthroughs. First, hardware's power increased and hardware's price decrease. As a consequence, new software emerged that takes advantage of this hardware by automating processes like load balancing and optimization across a huge cluster of nodes.

One of the problems with managing large quantities of data, has been the impact of latency that represents an issue in every aspect of computing, including communications, data management, system performance, and more. The capability to leverage distributed computing and parallel processing techniques reduced latency. It may not be possible to construct a big data application in a high latency environment if high performance is needed. It is necessary to process, analyse and verify this data in near real time. With the aim of reducing latency various distributed computing and parallel processing techniques have been proposed by researchers and practitioners from time to time.

Frequently problems are also related to high likelihood of hardware failure, improportionate distribution of data across various nodes in cluster and security issues due to the data access from anywhere.

The solution of those problems are typically based on distributed file storage (such as HDFS,[5] OpenAFS,[6] XtreemFS,[7]...), cluster resource management (such as YARN, Mesos,[8]...), and parallel programming model for large data sets and analysis model (such as MapReduce, Spark, Flink[9]).

The term Big Data is a broad and evolving term that refers to any collection of data so wide as to make it difficult or impossible to store it in a traditional software

---

[5] https://hadoop.apache.org/docs/r1.2.1/hdfs_design.html.

[6] https://www.openafs.org/.

[7] www.xtreemfs.org/.

[8] mesos.apache.org/.

[9] https://flink.apache.org/.

system, as RDBMS (Relational Database Management System). Although the term does not refer to any particular amount, usually it is possible to talk about Big Data from couple of Gigabytes of data, that is, when the data can not be easily processed by a single process. Big Data solutions are ideal for analysing not only raw structured data, but semistructured and unstructured data from a wide variety of sources [4]; Big Data solutions are ideal when all, or most, of the data needs to be analysed versus a sample of the data; or a sampling of data is not nearly as effective as a larger set of data from which to derive analysis; Big Data solutions are ideal for iterative and exploratory analysis when measures on data are not predetermined.

The collection of data streams of higher velocity and higher variety brings several problems that can be addressed by big data technologies. Thanks to big data technology it is possible to build an infrastructure that delivers low, predictable latency in both capturing data and in executing simple and complex queries; it is also possible to handle very high transaction volumes, often in a distributed environment; and supports flexible, dynamic data structures [10]. When dealing with such a high volume of information, it is relevant to organize data at its original storage location, thus saving both time and money by not moving around large volumes of data. The analysis may also be done in a distributed environment, where some data will stay where it was originally stored and be transparently accessed for required analytics such as statistical analysis and data mining, on a wider variety of data types stored in diverse systems; to scale for extreme data volumes and deliver faster response times. Most importantly, the infrastructure must be able to integrate analysis on the combination of big data and traditional enterprise data. New insight comes not just from analyzing new data, but from analyzing it within the context of the old to provide new perspectives on old problems [10]. Context-aware Big Data solutions could focus only on relevant information by keeping high probability of hit for all application-relevant events, with manifest advantages in terms of cost reduction and complexity decrease [11]. Obviously the results of big data analysis are only as good as the data being analyzed.

In last two decades, the term database is used in several contexts and is usually used as synonymous with SQL. Recently, however, the world of data storage has changed and new and interesting possibilities are now based on NoSQL. NoSQL stands for "Not Only SQL" and this emphasizes that the NoSQL technology is not entirely incompatible with SQL (Structured Query Language), it describes a large class of databases which are generally not queried with SQL. NoSQL data stores are designed to scale well horizontally and run on commodity hardware. NoSQL is definitely not suitable for all uses and is not a replacement of the traditional RDBMS database, but it can assist them or in part replace, and its main advantages make it useful, if not essential, in some occasions. NoSQL can significantly reduce development time because it eliminates the need to address complex SQL queries to extract structured data. The NoSQL database, if used properly, return the data in a timely way than a traditional database. This factor is really important with web and mobile applications. NoSQL data stores have several key features [12] that help them to horizontally scale throughput over many servers, replicate and distribute data over many servers, and dynamically add new attributes to data records [12]. NoSQL Data Models can be classified in:

- Key-value data stores (KVS). They store values associated with an index (key). KVS systems typically provide replication, versioning, locking, transactions, sorting, and/or other features. The client API offers simple operations including puts, gets, deletes, and key lookups.
- Document data stores (DDS). DDS typically store more complex data than KVS, allowing for nested values and dynamic attribute definitions at runtime. Unlike KVS, DDS generally support secondary indexes and multiple types of documents (objects) per database, as well as nested documents or lists.
- Extensible record data stores (ERDS). ERDS store extensible records, where default attributes (and their families) can be defined in a schema, but new attributes can be added per record. ERDS can partition extensible records both horizontally (per-row) or vertically (per-column) across a datastore, as well as simultaneously using both partitioning approaches.

Another important category is constituted by Graph data stores. They [13] are based on graph theory and use graph structures with nodes, edges, and properties to represent and store data. Key-Value, Document based and Extensible record categories aim at the entities decoupling to facilitate the data partitioning and have less overhead on read and write operations, whereas Graph-based category take the modeling the relations like principal objective. Therefore techniques to enhancing schema with a Graph-based database may not be the same as used with Key-Value and others. The graph data model fits better to model domain problems that can be represented by graph as ontologies, relationship, maps etc. Particular query languages allow querying the data bases by using classical graph operators as neighbour, path, distance etc.

Because for many Big Data use cases, the data does not have to be 100 percent consistent all the time, applications can scale out to a much greater extent. Eric Brewer's CAP theorem [14], formalized in [15], which basically states that is impossible for a distributed computing system to simultaneously provide all three of the following guarantees: Consistency, Availability and Partition Tolerance (from these properties the CAP acronym has been derived). Where:

- Consistency: all nodes see the same data at the same time
- Availability: a guarantee that every request receives a response about whether it was successful or failed
- Partition Tolerance: the system continues to operate despite arbitrary message loss or failure of part of the system that create a network partition

Only two of the CAP properties can be ensured at the same time. Therefore, only CA systems (consistent and highly available, but not partition-tolerant), CP systems (consistent and partition tolerant, but not highly available), and AP systems (highly available and partition-tolerant, but not consistent) are possible and for many people CA and CP are equivalent because loosing in Partitioning Tolerance means a lost of Availability when a partition takes place.

There are several other compute infrastructures to use in various domains. MapReduce is a programming model and an associated implementation for processing and generating large datasets. Users specify a map function that processes a key/value pair to generate a set of intermediate key/value pairs, and a reduce function that merges all intermediate values associated with the same intermediate key. Many real world tasks are expressible in this model, as show in [16]. Programs written in this functional style are automatically parallelized and executed on a large cluster of commodity machines. This allows programmers without any experience with parallel and distributed systems to utilize the resources of a large distributed system easily. Ather key concepts related to Big Data Analytics are:

Bulk synchronous parallel processing [17] is a model proposed originally by Leslie Valiant. In this model, processors execute independently on local data for a number of steps. They can also communicate with other processors while computing. But they all stop to synchronize at known points in the execution; these points are called barrier synchronization points. This method ensures that deadlock problems can be detected easily.

Large data streaming generated by thousands of data sources at high velocity, in high volume. It contains valuable potential insights and need to be processing real time to capture and pipe streaming data, but also to enrich, add context, personalize, and act on it before it becomes data at rest. These high-velocity applications require the ability to analyze and transact on streaming data.[10]

Large scale In memory computing, necessary to meet the strict real-time requirements for analyzing mass amounts of data and servicing requests within milliseconds an in-memory system/database that keeps the data in the random access memory (RAM) all the time [1].

High availability (HA) that is the ability of a system to remain up and running despite unforeseen failures, avoiding unplanned downtime or service disruption. HA is a critical feature that businesses rely on to support customer-facing applications and service level agreements.[11]

3   Key Aspects of Big Data Analytics

In recent years data, data management and the tools for data analysis have undergone a transformation. We have seen a significant increase in data collected by users thanks to web applications, sensors, etc. Unlike traditional systems, the type and the amount of data sources are varied. There is no longer just dealing with structured data, but also unstructured data from social networks, sensors, from the web, smartphones, etc. The acquisition of Big Data can be done in different ways, depending on the data source. The means for the acquisition of data can be divided into four categories: Application Programming Interface: the APIs are protocols used as a

---

[10] https://www.voltdb.com/fast-data.

[11] https://www.mapr.com/resources/high-availability-mapr.

communication interface between software components. Examples of APIs are the Twitter API, the Facebook Graph API and API offer by some search engines like Google, Bing and Yahoo! and the weather API. They allow, for example, to get the tweets related to specific topics (Twitter API) or examining the advertising content based on certain search criteria in the case of the Facebook Graph API. Web Scraping where data are simply taken by analysing the Web, i.e. the network of pages connected by hyperlinks. This has given rise to the term Big Data, that has become very popular, but its meaning often takes on different aspects. In general, we can summarize its meaning as a way to treat large volumes of data constantly increasing [7], an action that requires instruments for collecting, storage and analysis different from the traditional ones. In particular we refer to datasets that are so large to be not manageable by traditional systems, such as relational DBMS running on a single machine. In fact, when the size of a dataset is more than few terabytes, it is necessary to use a distributed system, in which the data is partitioned across multiple machines. Several technologies to manage Big Data have been created that are able to use the computing power and the storage capacity of a cluster, with an increase in performance proportional to the number of machines present on the same cluster. Those technologies provide a system for storing and analysing distributed data. Using redundancy of data and sophisticated algorithms, can work even in the event of failure of one or more machines in the cluster, transparently to the user. Distributed systems provide the basis for those systems. In fact a distributed architecture is able to serve as an umbrella for many different systems.

4 Popular Technologies for Big Data Analytics Utilizing Concepts of Distributed Computing

In the subsections below we discuss few popular open source Big Data technologies those are wideliy used to day across various industries.

4.1 Hadoop

The Hadoop Distributed File System (HDFS) [18] is a distributed filesystem written in Java designed to be run on commodity hardware, in which the data stored are partitioned and replicated on the nodes of a cluster. HDFS is fault-tolerant and developed to be deployed on low-cost machines. Hadoop is just one example of a framework that can bring together a broad array of tools such as (according to Apache.org): Hadoop Distributed File System that provides high-throughput access to application data; Hadoop YARN for job scheduling and cluster resource management; Hadoop MapReduce for parallel processing of big data. Hadoop, for many years, was the leading open source Big Data framework but recently the newer and more advanced

Spark has become the more popular of the two Apache Software Foundation tools. Hadoop can run different applications, including MapReduce, Hive and Apache Spark. Through redundancy of data and sophisticated algorithms, Hadoop can work even in the event of failure of one or more machines in the cluster, transparently to the user. Hadoop is an open-source software system used extensively in this area, offering both a distributed file system for storing information that one for their computing platform. The module supports multiple software for the analysis of data, including MapReduce and Spark. The substantial difference between these two systems is that MapReduce obliges to store the data to disk after each iteration, while Spark can work in main memory, exploiting the disc only in case of need. The Spark system, which is a high-level framework, provides a set of specific modules for each scope.

4.2 Yarn

YARN (Yet Another Resource Negotiator) is a main feature of the second version of Hadoop. Before YARN, the same node of the cluster, on which he was running the Job Tracker, took care of both of the cluster resource management is the scheduling of the task of MapReduce applications (which were the only possible ones). With the advent of YARN the two tasks were separated and were held respectively by the ResourceManager and AppliationMaster.

4.3 Hadoop Map Reduce

Hadoop MapReduce is a programming model for processing large data sets on parallel computing systems. A MapReduce Job is defined by: the input data; a procedure Map, which for each input element generates a number of key / value pairs; a phase of shuffle network; It reduces a procedure, which receives as input elements with the same key and generates a summary information from such elements; the output data MapReduce guarantees that all elements with the same key will be tried by the same reducer, since the mapper all use the same hash function to decide which reducer send the key / value pairs.

4.4 Spark

Apache Spark is a project that otherwise to Hadoop MapReduce does not require the use of your hard disk, but may enter directly into the main memory managing to offer performance even 100 times on specific applications. Spark offers a broader set of primitive compared to MapReduce, greatly simplifying programming.

5 Conclusion

A distributed computing system consists of number of processing elements inter-connected by a computer network and co-operating in performing certain assigned tasks. When data becomes large, the database is distributed into various sites. The distributed databases need distributed computing to store, retrieve, and update data in a well coordinated way [9]. The advent of Big Data has led in recent years in search of new solutions for storing them and for their analysis. To manage Big Data, technologies have been created that are able to use the computing power and the storage capacity of a cluster, with an increase in performance proportional to the number of machines present on the same. In particular big data analytics is a prom-ising area for next generation of innovation in the field of automation, with the ever increasing need of extracting value from data in several field of application. With that objetcive in mind various technologies/system have been evolved in last decade or so. The most used of these systems is Hadoop, which provides a system for stor-ing and analyzing distributed data. YARN is a main feature of the second version of Hadoop, born to solve common problems. Hadoop Map Reduce, is designed for processing large data sets with a parallel and distributed algorithm on a cluster, and Spark performs in-memory processing of data. In this chapter an overview of tech-nologies for Big Data analytics based on Distributed Computing concepts have been presented. With the increasing amount of data, the analytics will be ever more important in the decision-making process in several sectors allowing the discovery of new opportunities and increasing the quality of information.

References

1. Gartner. Hype cycle for big data, 2012. Technical report (2012) On the role of Distributed Computing in Big Data Analytics 11
2. Afgan, E., Bangalore, P., Skala, K. Application information services for distributed computing environments. Future Generation Computer Systems 27 (2011) 173–181
3. Cattell, R. Scalable sql and nosql data stores. Technical report (2012)
4. Brewer, E.A. Towards robust distributed systems (abstract). In: Proceedings of the nineteenth annual ACM symposium on Principles of distributed computing. PODC '00, New York, NY, USA, ACM (2000) 7-.
5. Nessi: Nessi white paper on big data. Technical report (2012)
6. Dean, J., Ghemawat, S. Mapreduce: simplified data processing on large clusters. In: Osdi04: Proceedings Of The 6th Conference On Symposium On Operating Systems Design And Implementation, Usenix Association (2004)
7. IBM, Zikopoulos, P., Eaton, C. Understanding Big Data: Analytics for Enterprise Class Hadoop and Streaming Data. 1st edn. McGraw-Hill Osborne Media (2011)
8. Schroeck, M., Shockley, R., Smart, J., Romero-Morales, D., Tufano, P. Analytics: The real-world use of big data. Ibm institute for business value – executive report, IBM Institute for Business Value (2012)
9. Gilbert, S., Lynch, N. Brewer's conjecture and the feasibility of consistent, available, partition-tolerant web services. SIGACT News 33 (2002) 51–59

10. Zhang, H., Chen, G., Ooi, B.C., Tan, K.L., Zhang, M. In-memory big data management and processing: A survey. IEEE Transactions on Knowledge and Data Engineering 27 (2015) 1920–1948

11. Valiant, L.G. A bridging model for parallel computation. Commun. ACM 33 (1990) 103–111

12. Oracle: Big data for the enterprise. Technical report (2013)

13. Robinson, I., Webber, J., Eifrem, E. Graph Databases. O'Reilly Media, Incorporated (2013)

14. White, T. Hadoop: The Definitive Guide. 1st edn. O'Reilly Media, Inc. (2009)

15. Grover, P., Johari, R. Bcd: Bigdata, cloud computing and distributed computing. In: Communication Technologies (GCCT), 2015 Global Conference on, IEEE (2015) 772–776

16. Gartner: Pattern-based strategy: Getting value from big data. Technical report (2011)

17. Gandomi, A., Haider, M. Beyond the hype: Big data concepts, methods, and analytics. International Journal of Information Management 35 (2015) 137–144

18. Amato, A., Venticinque, S. In: Big Data Management Systems for the Exploitation of Pervasive Environments. Springer International Publishing, Cham (2014) 67–89

19. Afgan, E., Bangalore, P., Skala, T. Scheduling and planning job execution of loosely coupled applications. The Journal of Supercomputing 59 (2012) 1431–1454

Fundamental Concepts of Distributed Computing Used in Big Data Analytics

Qi Jun Wang

1 Introduction

The study of distributed computing became its own branch of computer science in the late 1970s and early 1980s. So it has been a long time since the advent of distributed computing technology and since then many fundamental concepts of Distributed Computing has been successfully used in various areas of real life applications. These fundamental concepts are the keys to achieve large-scale computation in a scalable and affordable way and hence most of the Big Data Technologies of today leverage those concepts to design their internal frameworks and features. In turn those Big Data Technologies are used to build applications around Big Data Analytics for various industries.

In this chapter we provide detail understanding of some of these fundamental concepts that are must to know by any Big Data Analytics practitioner. We also provide appropriate examples around these concepts wherever necessary. We start with explanation of the concepts of Multi-threading and Multi processing. Next we introduce the different types of computer architecture along with the concepts of scale up and scale out. Next we delve into the principles of Queuing system and use of the same in Distributed Computing. We also cover the relationship between Consistency, Availability, and Partition Tolerance and their trade of in Cap Theorem. Next we provide the concept of Computing Cluster and main challenges in the same. Finally we end with discussion around key Quality of Service (QoS) requirements applicable in Big Data Analytics area.

Q.J. Wang (✉)
Big Data Architect, Lab Service, IBM Analytics, Seattle, Washington, DC, USA
e-mail: wangqij@us.ibm.com

© Springer International Publishing AG 2017
S. Mazumder et al. (eds.), *Distributed Computing in Big Data Analytics*,
Scalable Computing and Communications, DOI 10.1007/978-3-319-59834-5_2

2 Multithreading and Multiprocessing

Multi-threading and Multi processing are two fundamental concepts in Distributed Computing. They are widely used to enhance the performance of Distributed Computing system. The main purpose of Multi threading and Multi processing is to enhance the parallelization, which reduces the system process delay.

2.1 Concept of Multiprocessing

Multiprocessing is a mode of operation in which two or more processors in a computer simultaneously process two or more different portions of the same program (set of instructions). Supercomputers typically combine thousands of such microprocessors to interpret and execute instructions. The advantage of multiprocessing is it can dramatically enhance the system throughput and speed up the execution of programs.

2.2 Example of Multiprocessing

The concept of multiprocessing has been used in many famous distributed computing or big data platform, such as Apache Hadoop. In Hadoop, users can concurrently start multiple mappers and reducers and each mapper or reducer corresponds to one process.

Figure 1 is the picture showing the multiprocessing model in the Hadoop runtime environment:

Hadoop client is responsible for submitting map-reduce jobs to the resource manager, and resource manager will look up the available resources (CPU, memory) on each slave node and allocate these resources to the Hadoop applications. After that, Hadoop application will split the jobs and start concurrent multi processes (mappers) to process each splits. Finally, it will start another set of concurrent multi processes (reducers) to combine the results of mappers and output data to Hadoop Distributed File System (HDFS).

2.3 Concept of Multithreading

A thread is the smallest sequence of programmed instructions that can be managed independently by a scheduler. Multithreading is the ability of a central process unit (CPU) or a single core in a multi-core processor to execute multiple threads concurrently, appropriately supported by the operating system.

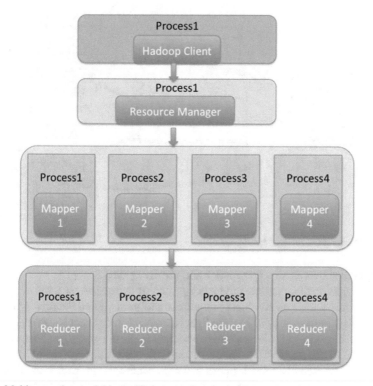

Fig. 1 Multiprocessing model in the Hadoop runtime environment

Multithreading aims to increase utilization of a single core by using thread-level as well as instruction-level parallelism, and the advantage of Multithreading is If a thread gets a lot of cache misses, which is s a state where the data requested for processing by a component or application is not found in the memory, the other threads can continue taking advantage of the unused computing resources, like CPU and memory. Also, if a thread cannot use all the computing resources of the CPU (because instructions depend on each other's result), running another thread may prevent those resources from becoming idle [2]. If several threads work on the same set of data, they can actually share their cache, leading to better cache usage or synchronization on its values.

2.4   *Example of Multithreading*

Apache Spark is one of the typical big data platforms using multi threading. Spark implements based on multithreading models for lower overhead of JVM (Java Virtual Machine) and data shuffling between tasks.

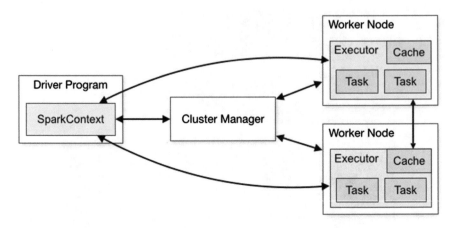

Fig. 2 Apache spark multithreading model

Figure 2 shows the Apache spark multi threading model:

Spark applications run as independent sets of processes on a cluster, coordinated by the SparkContext object in the main program (called the driver program). Specifically, to run on a cluster, the SparkContext can connect to several types of cluster managers (either Spark's own standalone cluster manager, Mesos [20] or YARN [21] (Yet Another Resource Manager)), which allocate resources across applications. Once connected, Spark acquires executors on machines in the cluster, which are processes that run computations and store data for your application. Next, it sends your application code (defined by JAR or Python files passed to SparkContext) to the executors. Finally, SparkContext sends tasks to the executors to run. Each application gets its own executor processes, which stay up for the duration of the whole application and run tasks in multiple threads. So, we can see that each executor is a process, but it includes multi threading (Task) to run the application.

2.5 *Difference between Multiprocessing and Multithreading*

A process is an executing instance of an application and it has a self-contained execution environment. A process generally has a complete, private set of basic run-time resources; in particular, each process has its own memory space. Also, a process can contain multiple threads.

A thread is a basic unit of CPU utilisation; it comprises a thread ID, a program counter, register set, and a stack. It shared with other threads belonging to the same process its code section, data section and other operating system resources such as open files and signals. A thread of execution is the smallest sequence of programmed instructions that can be managed independently by a scheduler, which is typically a part of the operating system.

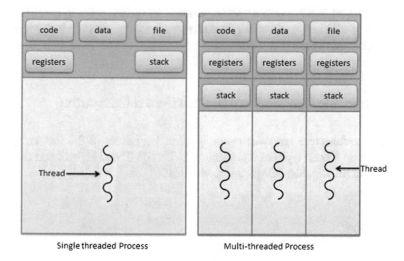

Single threaded Process   Multi-threaded Process

Fig. 3 Difference between process and thread [3]

Figure 3 is the picture showing the difference between process and thread:

From above picture, you can see typically one process can have one or multi threads and all the threads in one process share the same code, data and files, but they have independent registers and stack.

It's important to note that a thread can do anything a process can do. But since a process can consist of multiple threads, a thread could be considered a 'lightweight' process, like short-lived request to a web application for getting a user details. Thus, the essential difference between a thread and a process is the work that each one is used to accomplish. Threads are used for small tasks, whereas processes are used for more 'heavyweight' tasks, like a batch ETL job.

In addition, threads can share data among them, which processes cannot and hence they can communicate easily, Threads take lesser time to get started compared to processes and through Threads multiple user requests can be supported concurrently.

The implementation of threads and process differs between operating systems, but in most cases a thread is a component of a process. Multiple threads can exist within one process, executing concurrently and sharing resources such as memory and open files, while different processes do not share these resources. In particular, the threads of a process share its executable code and the values of its variables at any given time.

Threads may not be actually running in parallel. It is the operating system, which does intelligent multiplexing so that the shares of the processes provided to each thread in a manner that it appears like the threads are executed in parallel.

In summary, multithreading and multiprocessing are two basic technologies to improve the system throughput, and as multicore computers are becoming more and more prevalent, a large number of distributed computing platform now support multithreading and multiprocessing. Big Data Technologies, like Spark, Hadoop, etc. use the Multithreading and Multiprocessing in various ways to ensure speedy execution of different types of Big Data Analytics jobs so that the insights can be created within an acceptable timeframe.

3   Computing Architecture in Distributed Computing

Computer architecture has been evolving since the advent of the first computer. Now there are 3 main types of architecture: SISD, SIMD and MIMD, and there are two types in MIMD: SM-MIMD and DM-MIMD.

3.1   SISD

At the very beginning, most of the computers used scalar processors, whose instructions operate on single data. Such processor model was called SISD (Single Instruction Single Data). It is very slow as there is no parallelism in such model.

3.2   Vector Processor

Vector processor, also known as array processor was invented in the 1970s, which implements an instruction set containing instructions that operate on one-dimensional arrays of data called vectors. Vector processors can greatly improve performance on certain workloads, such as arithmetical operation and digital signal processing. Today most commodity CPUs implement architectures that feature instructions for a form of vector processing on multiple data sets. Meanwhile, many companies, like Intel and IBM, provide Vector Processing library for users to develop their own Vector Processing program.

There are two types of vector processing: SIMD (Single Instruction Multiple Data) and MIMD (Multiple Instruction Multiple Data). They both provide data processing parallelism, and the difference is SIMD only provide the data level parallelism while MIMD can provide two dimensional parallelism: instruction level and data level.

3.3   SIMD

SIMD is widely used for graphics and video processing, vector processing and digital signal processing. It is short for Single Instruction Multiple Data, which is one classification of computer architectures. SIMD operations perform the same computation on multiple data points resulting in data level parallelism and thus performance gains.

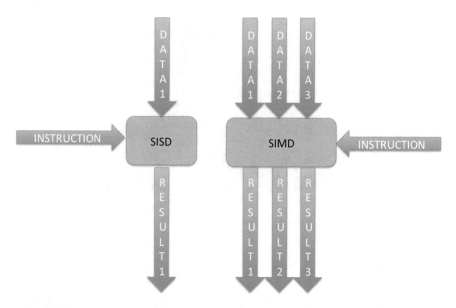

Fig. 4 Difference between SISD and SIMD

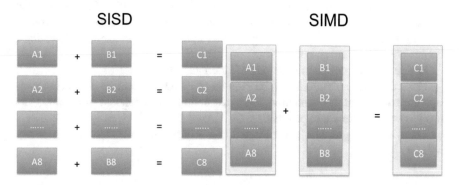

Fig. 5 SISD and SIMD example

Figure 4 is the picture to show what's the difference between SISD and SIMD:

It can be seen from the picture that SIMD doesn't provide instruction level parallelism, but only data level parallelism. It can process multiple data vectors with one instruction. This is very useful for some loop operation. For example, if you have two Byte lists and you want to add them to one list, assuming the length of the two lists is 1024, then it will take 1024 times to complete the adding operation, but if SIMD is supported by the computer and the CPU is 64-bits, it will only take 128 times to finish the processing.

Figure 5 is the picture to show this example:

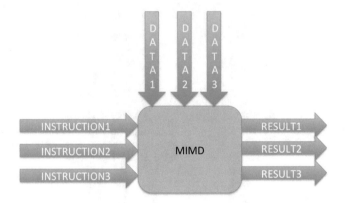

Fig. 6 MIMD parallelism

3.4 MIMD

MIMD (Multiple Instruction Multiple Data) is another type of parallelism. Compared with machine with SIMD, machines using MIMD have a number of processors that function asynchronously and independently, [4] which means that parallel units have separate instructions, so each of them can do something different at any given time; one may be adding, another multiplying, yet another evaluating a branch condition, and so on.

Figure 6 is the picture to show MIMD parallelism:

From the above picture, it can be seen that MIMD architecture can accept multiple instructions at the same time. Each instruction is independent from others and has its own data stream to process.

There are two types of MIMD: Shared-Memory MIMD and Distributed-Memory MIMD.

3.5 SM-MIMD

In the Shared-Memory (SM) Model, all the processors share a common, central memory. The distinguishing feature of shared memory systems is that no matter how many memory blocks are used in them and how these memory blocks are connected to the processors, address spaces of these memory blocks are unified into a global address space, which is completely visible to all processors of the shared memory system [5].

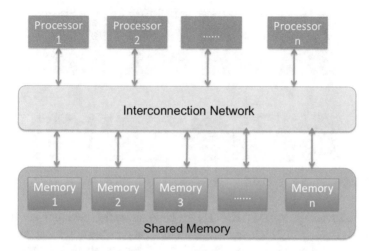

Fig. 7 Shared memory MIMD

Figure 7 is the SM-MIMD picture showing processors and memories are connected by interconnection network:

One of the advantages of Shared-Memory model is it is easy to understand and another advantage is that memory coherence is managed by the operating system and not the written program, so it is easy for developer to design parallel program in such model. The disadvantage is that it is difficult to scale out with Shared-Memory model and it is not as flexible as Distributed-Memory model.

3.6   DM-MIMD

Distributed-Memory (DM) is another type of MIMD. In this model, each processor has its own individual memory location. Each processor has no direct knowledge about other processor's memory. For data to be shared, it must be passed from one processor to another as a message. Since there is no shared memory, contention is not as great a problem with these machines [4].

DM-MIMD is the fastest growing part in the family of high performance computers or servers as it can dramatically enhance the bandwidth by adding more processors and memories.

Figure 8 is the picture showing the structure of DM-MIMD:

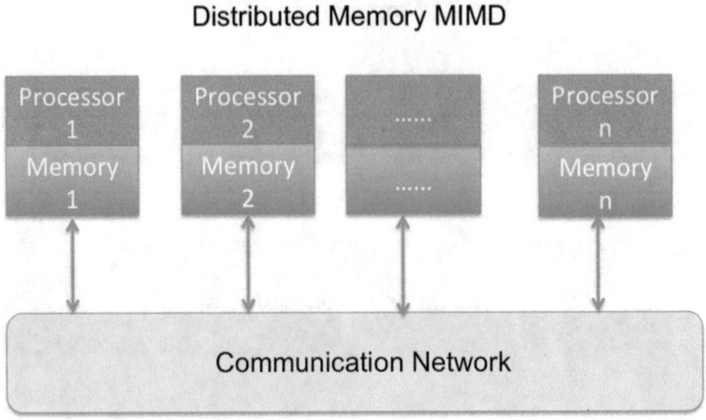

Distributed Memory MIMD

Fig. 8 Distributed memory MIMD

The disadvantage of DM-MIMD is the communication cost between different processors can be very high and it is difficult to access the non-local data, which is located in other processors' memories. Nowadays, there are many system designs to reduce the time and difficulty between processors, like Hypercube and Mesh.

MPP (massively parallel processors) is one of the typical examples of DM-MIMD and many famous big data technologies are base on MPP, like BIG SQL (SQL on Hadoop) from IBM and Impala from Cloudera.

In summary, MIMD is a trend in current computer architecture development and most of the distributed computing systems are based on such technologies.

4 Scalability in Distributing Computing

Scalability is a frequently mentioned concept in Distributed Computing area. It means the capability of a system to handle a growing amount of work, or its potential to be enlarged in order to accommodate that growth. In this section, it will cover the definition of scalability, comparison of scale up method and scale out method.

4.1 Scalability Requirement and Category

In the Internet era, rapid data growth is happening every day and such growth is bringing a lot of challenges to most of business and industries. As a result, every organization today has a need to build or design systems with reasonable scalability characteristic.

There are two approaches related to scalability: scale up and scale out. They are commonly used in discussing different strategies for adding functionality to

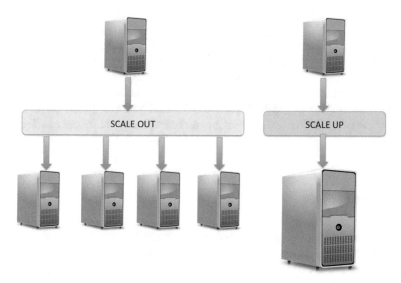

Fig. 9 Basic difference of scale up and scale out

hardware systems. They are fundamentally different ways of addressing the need for more processor capacity, memory and other resources.

Figure 9 is the picture showing the basic difference of scale up and scale out.

4.2 Scaling Up

Scaling up, also known as vertical scaling, means upgrading hardware. It generally refers to purchasing and installing a more capable central control or piece of hardware. For example, when a application's data demands start to push against the limits of an individual server, a scaling up approach would be to buy a more capable server with more processing capacity and RAM [6].

The advantages of scale up are:

- Availability of high amount memory can help processing lots of data with low latency.
- It is easier to control as you only upgrade the hardware, like CPU, memory, network, disk in the same machine
- Less power consumption than running multiple servers as there are less machines in the scale up methodology.
- Less cooling cost in the data center.

The disadvantage of scale up is as follows:

- High price of the high-performance servers. Typically, scale up can be more expensive as you have to buy a lot of powerful hardware (CPU, Memory, Disk) and such hardware is much more pricy than ordinary one.
- Furthermore, sometimes scale up is not regarded as feasible because of the data explosion and the unmatched limits to individual hardware pieces on the market.
- In terms of fault tolerance, there is greater risk of hardware failure causing bigger outages.

4.3   Scaling Out

By contrast, scaling out, also known as horizontal Scaling, means adding many lower-performance machines to the existing system to extend the computing resource and storage capacity [6]. With these types of distributed setups, it's easy to handle a bigger data volume by running data processing across the whole system, which may include thousands of lower-performance machines.

Scale out has been gaining more and more popularities these days. Scale out architecture started getting popular when web applications supporting 100 s of users concurrently became popular in early 2000. The benefits of scale out methodology are:

- It is easy to add more storage and computing resource to the existing system by adding some low-performance computers.
- Another advantage is the price. Usually, the cost of scale out system is much lower than scale up system as most ordinary computers are much cheaper than high-performance computers.
- Most importantly, scale out provides a true scalability, which means the system capacity can be extend to an unlimited level by adding more computers to the system.
- In terms of fault tolerance, scale out is also easier as typically there is mechanism inside scale out system, which will put some standby nodes or servers to particular service and make data replication across the servers or even racks in the data center. Such mechanism makes it very easy to recover the service and data.

The disadvantages of scale out system are:

- The maintenance of such a big platform. It may take several days to trace one problem because it is very difficult to judge which node causes the problem and where is the log.
- Another drawback is in data center scale out system will take up more space, so the electricity and cooling expense are more expensive than scale up system.

4.4   Prospect of Scale Up and Scale Out

Nowadays Scale up and scale out are both growing rapidly. On the one hand, some companies, like IBM, Intel are still investing large amount of money on the advanced high-performance computer research and development that can support scale up. For example, IBM recently announced the latest POWER9 chip, which has up to 24 cores and provides blazing throughput to speed up complex calculations. On the other hand, most of the Internet companies, like Google, Facebook and Yahoo invest a lot on the scale out system development. Apache Hadoop is one of the most successful projects in the scale out area. In Hadoop, users can easily extend the storage size and computing resource by adding new nodes to the existing system.

However, scale up and scale out are not mutual exclusive. There are many cases where scale up and scale out are going hand in hand. For instance, in some data centers, adding a large number of new servers happens together with the upgrading of old servers, like more CPUs, more memory and more disks.

For example, in many real life Big Data Analytics systems, where the data growth is very fast and the big data cluster cannot process the high volume of data within the expected timeframe, both scale up and scale out approaches are leveraged. The specific measures taken are

- Put more memory in the existing servers to make the data analytics faster, which is scale up
- Add more servers to the cluster to extend the volume of the storage, which is scale out

In a nutshell, scalability is one of most important features of distributed computing system. Scale up and scale out are two main technologies to address the scalability problem. These two methods are in nature different and designed to be used in different scenarios. Typucal systems supporting Big Data Analytics leverage both of these approaches optimaly as needed to address the scalability concerns of specific cases.

5   Queuing Network Model for Distributed Computing

Queue system and Queue network model are mainly used to describe and analyze the quality of service in distributing computing system, and it is the theoretical basis of service scheduling in big data area. In this section, some basic characters of queue system will be presented.

5.1 Asynchronous Communication

Asynchronous communication is the basic concept behind the Queuing technology. Synchronous communication is occurring in real time, like a phone call. You have to wait until the person on the other end answers your question in real time. When you are using asynchronous communication, you are not waiting for a response in real time. You can move on to another task before your first task is completely finished or once you are done with your part of a task. Email is a good example of asynchronous messaging. As soon as the email is sent from you, you can continue handling other things without the need of getting an immediate response from the receiver [23]. You can do other things while the message is in transit.

For example, if a web application receives a lot of requests, the Asynchronous Communication mechanism will let this web application generate tasks in response to user inputs, and then tasks will be sent to a receiver. A receiver can retrieve the task and process it when the receiver is ready and return a response when it is finished. In this a way the user interface can remain responsive all the time.

5.2 Queue System

Queue system is based on the asynchronous communication. A queuing system consists of one or more servers that provide service of some sort to arriving customers [7]. The customers represent workloads, users, jobs, transactions or programs. Customers who arrive to find all servers busy generally join one or more queues (lines) in front of the servers, and leave the system after being served.

Figure 10 shows how a typical queuing system works.

Typically, A queuing system is characterized by following components: distribution of inter-arrival times, distribution of service times, the number of servers, the service discipline and the maximum capacity [8]. There are several everyday examples that can be described as queuing systems, such as bank-teller service, computer systems, manufacturing systems, maintenance systems, communications systems and so on.

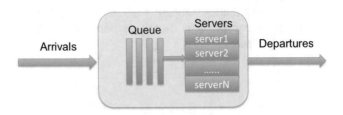

Fig. 10 Queuing system model

5.3 Queue Modeling

Queuing modeling is an analytical modeling technique for the mathematical analysis of systems with waiting lines and service stations. In queuing modeling, a model is constructed so that queue lengths and waiting time can be predicted.

There are two types of queuing: Single queuing service and Queuing Network.

A single queuing service consists of one or more identical servers with a joint waiting room. Jobs arrive at the queue with an arrival rate and have an expected service time. If the servers are all occupied, jobs have to line up in the queue. After being served, jobs will leave the queue.

A Queuing Network Model consists of a number of interconnected queues, which are connected by customer routing. After a customer is serviced at one node, it can join another node and queue for service, or leave the network directly.

Queuing networks can be classified into three categories: open, closed, and mixed queuing networks. Open queuing networks have an external input and an external final destination. In closed queuing networks the customers circulate continually never leaving the network. Mixed queuing networks combine open and closed Queuing, which means Open for some workloads and closed for others.

Queuing Network Models are now widely used to analyze computer system, communication system and product system. In the Distributing Computing area, Queuing Network Models can be used to analyze the workloads or jobs scheduling efficiency, such as the average waiting time, service processing time and throughput.

Typically, users can submit multiple jobs into distributed cluster. At first, scheduler will gather all the available resources, such as Idle CPU, memory in the distributed cluster. If there are enough resources in the cluster, all the jobs can be executed concurrently and then all the jobs leave the cluster after being served. If the resources in the cluster in not enough, all the jobs will be put in one or multi queues and they have to wait for the scheduler to run the jobs one by one. Usually, there are different strategies to schedule jobs, such as FIFO (first input first out), LIFO (last input first out) and Priority based method. Different services may adopt different strategies and some of them can support user-defined strategies. For some types of service, they can set different priorities for the different queues, and users can submit jobs to different queues according to the job processing time and job priorities.

The technologies popularly used to achieve asynchronous communication/queuing in Big Data Analytics world are Yarn, Mesos, Kafka, etc. The fundamental unit of scheduling in YARN and Mesos is a queue. The capacity of each queue specifies the percentage of cluster resources that are available for applications submitted to the queue. Queues can be set up in a hierarchy that reflects the database structure, resource requirements, and access restrictions required by the various organizations, groups, and users that utilize cluster resources. On the other hand, Kafka provides implementation of application level Queue where actual applications can send some tasks/messages that can be asynchronously acted upon by other applications.

Fig. 11 CAP theorem [19]

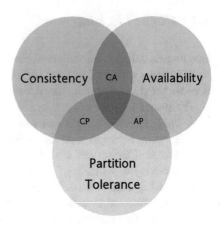

In summary, queue network modeling provides a methodology to analyze the service quality and then improve the service quality based on the analyze result.

6   Application of CAP Theorem

CAP theorem is very famous in distributed computing system. The CAP Theorem, also known as Brewer's theorem, states that, in a distributed system (a collection of interconnected nodes that share data.), you can only have two out of the following three guaranteed across a write/read pair: Consistency, Availability, and Partition Tolerance – one of them must be sacrificed [10].

6.1   Basic Concepts of Consistency, Availability, and Partition Tolerance

Below is the detailed explanation of Consistency, Availability, and Partition Tolerance:

- Consistency – A read is guaranteed to return the most recent write for a given client.
- Availability – A non-failing node will return a reasonable response within a reasonable amount of time (no error or timeout).
- Partition Tolerance – The system will continue to function when network partitions occur [10].

Figure 11 shows the CAP theorem.

6.2 Combination of Consistency, Availability, and Partition Tolerance

According to CAP theorem, it is impossible to build a general data store that is continually available, sequentially consistent and tolerant to any partition pattern. You can build one that has any two of these three properties. All the combinations available are:

- CA – data is consistent between all nodes – as long as all nodes are online – and you can read/write from any node and the data is the same, but if you ever develop a partition between nodes, the data will be out of sync (and won't re-sync once the partition is resolved).
- CP – data is consistent between all nodes, and maintains partition tolerance (preventing data de-sync) by becoming unavailable when a node goes down.
- AP – nodes remain online even if they can't communicate with each other and will re-sync data once the partition is resolved, but you aren't guaranteed that all nodes will have the same data (either during or after the partition) [11]

No distributed system is safe from network failures, thus network partitioning generally has to be tolerated. In the presence of a partition, one is then left with two options: consistency or availability [12].

If a system chooses to provide Consistency over Availability in the presence of partitions, it will preserve the guarantees of its atomic reads and writes by refusing to respond to some requests. It may decide to shut down entirely (like the clients of a single-node data store), refuse writes (like Two-Phase Commit), or only respond to reads and writes for pieces of data whose master node is inside the partition component. There are plenty of things, which are made much easier (or even possible) by strongly consistent systems. They are a perfectly valid type of tool for satisfying a particular set of business requirements [13]. Typically, Database systems designed with traditional ACID (Atomicity, Consistency, Isolation, Durability) guarantees in mind such as RDBMS (relational database management system) choose consistency over availability [12].

If a system chooses to provide Availability over Consistency in the presence of partitions, it will respond to all requests, potentially returning stale reads and accepting conflicting writes. These inconsistencies are often resolved via causal ordering mechanisms like vector clocks and application-specific conflict resolution procedures. There are plenty of data models which are amenable to conflict resolution and for which stale reads are acceptable [13]. Systems designed around the BASE (Basically available, soft state, eventually consistent) philosophy, common in the No-SQL movement for example, choose availability over consistency [12].

In the absence of network failure, which means the distributed system is running normally, both availability and consistency can be satisfied. CAP is frequently misunderstood as if one had to choose to abandon one of the three guarantees at all times. In fact, the choice is really between consistency and availability for when a partition happens only; at all other times, no trade-off has to be made [12].

One of the typical AP systems is Apache Cassandra Database, in which availability and partition tolerance are generally considered to be more important than consistency in Cassandra. But Cassandra can be tuned with replication factor and consistency level to also meet C.

7 Quality of Service (QoS) Requirements in Big Data Analytics

In big data analytics area, there are many factors regarding to the Quality of Service (QoS) requirements, such as performance, Interoperability, fault-tolerance, Security, Manageability, Load-Balance, High-Availability and SLA.

7.1 Performance

Most of Distributed Computing systems are designed to enhance the Performance of computing or IO (input, output) speed, so Performance is one of the key QoS requirements. Typically 3 things are related to performance: Throughput (in terms of data), Response Time and support for concurrent Requests. What is important for many Big Data Analytics application is all three – like real Time Analytics which is accessed by 100 s of concurrent users and which needs to process large volume of data. Many advanced technologies can enhance the performance, like pre-computing, in memory processing, Thread level parallelism, using of hybrid storage like SSD + HDD etc.

In the cognitive computing area of Big Data Analytics, two types of advanced hardware technologies, FPGA (Field Programmable Gate Array) and GPU (graphics processing unit) are leveraged to accelerate the speed of machine learning model training and real time classification or prediction.

7.2 Interoperability

Interoperability is another important QoS requirement in Big Data ecosystem. It is the property that allows for the unrestricted sharing of resources between different systems. This can refer to the ability to share data between different components or machines, both via software and hardware, or it can be defined as the exchange of information and resources between different computers through local area networks (LANs) or wide area networks (WANs). Broadly speaking, interoperability is the ability of two or more components or systems to exchange information and to use the information that has been exchanged [17]. Interoperability is a very important feature as sharing data between different systems is inevitable in the big data era, so most big data technologies support interoperability.

For instance, some web applications provide many interfaces or API to access different databases or big data storage. Apache Zeppelin [22] and Jupyter Notebooks are widely used tools for exploration in Big Data Analytics which provide interoperability for accessing various data sources and sinks in a transparent manner.

7.3   Fault-Tolerance

An important challenge faced by today's big data analytics systems is fault-tolerance. It is very normal that when running a parallel query at large scale, some form of failure is likely to occur during execution. Fault tolerance is the property that enables a system to continue operating properly in the event of the failure of some of its components. Fault tolerance places a significant role in big data area as both cluster scale and data are becoming increasingly complicated. Typically, there are two types of failure when running big data application: data failure and node failure. Data failure means some intermediate partitions of data may be lost due to application design or hardware problem. Big data system should design the mechanism to handle such failure automatically.

Apache Cassandra is an open-source distributed NoSQL database management system and it is a good example of such mechanism. Apache Cassandra is not driven by a typical master-slave architecture, where failure of the master becomes a single point of system breakdown. Instead, it harbors a concept of operating in a ring mode so that there is no single point of failure. Whenever required, users can restart the nodes without the dread of bringing the whole cluster down.

Another real example of Fault-tolerance is that one application used checkpoint approach in the spark-streaming project. Figure 12 shows the Steaming process in this case.

In this case, the application set checkpoint in each time interval, so when job failure happens due to software, hardware or network problem, it can easily find the broken point and then restart the streaming process.

7.4   Security

Security is necessary in all Big Data Analytics systems. The big data explosion has given rise to a host of information technology tools and capabilities that enable organizations to capture, manage and analyze large sets of structured and unstructured data for actionable insights and competitive advantage. But with this new technology comes the challenge of keeping sensitive information private and

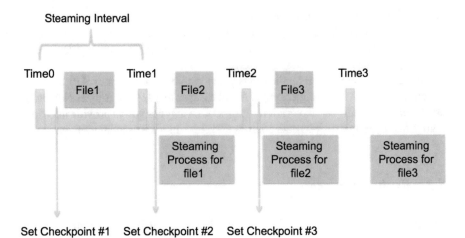

Fig. 12 Checkpoint in spark streaming

secure. Big data that resides within a big data environment can contain sensitive financial data in the form of credit card and bank account numbers. It may also containproprietary corporate information and personally identifiable information (PII) such as the names, addresses and social security numbers of clients, customers and employees. Due to the sensitive nature of all of this data and the damage that can be done should it fall into the wrong hands, it is imperative that it be protected from unauthorized access [18]. To handle security problem in big data environment, following aspects should be taken into consideration:

- Ensure the proper authentication of users who access the big data environment.
- Ensure that authorized users can only access the data that they are entitled to access.
- Ensure that data access histories for all users are recorded in accordance with compliance regulations and for other important purposes.
- Ensure the protection of data—both at rest and in transit—through enterprise-grade encryption [18].

Kerberos is a very popular service level securities tool in big data area. It is a network authentication protocol, and designed to provide strong authentication for client/server applications by using secret-key cryptography.

7.5 Manageability

Manageability is an indispensable requirement of big data analytics system to make the environment and services easily manageable. As big data systems are becoming increasingly complex, it is very important to provide system administrators and users with enough and user-friendly interface, which can facilitate the daily management, such as service installation and configuration, service start and stop, service status check, metrics collection and visualization, job history, service and job log.

Most of big data platforms provide good Manageability, such as Apache Hadoop. Hadoop is an ecosystem, not a single product, so there are many tools providing Hadoop service management and one of the outstanding ones is called Ambari.

7.6 Load-Balance

Load-Balance is a configuration in which cluster-nodes share computational workload to provide better overall performance. For example, a web server cluster may assign different queries to different nodes, so the overall response time will be optimized. However, approaches to load balancing may significantly differ among applications. For example, a high-performance cluster used for scientific computations would balance load with different algorithms from a web-server cluster, which may just use a simple round-robin method by assigning each new request to a different node [15].

In some popular Distributed Computing systems, like Apache Hadoop, Load-Balance is a very important feature. In Hadoop, Load balancing issues occur if there are some tasks significantly larger than others such that in the end only a few tasks are running while all others are finished. This situation happens in case of skewed reduce keys and can be easily identified (all tasks finished but a few). But the real challenge is not to detect load balancing issues but to either avoid data skew in the beginning (by clever partitioning and choice of parallelism) or to have adaptive methods that can mitigate the effect of data skew. Therefore, at first during the stage of job partitioning, it is critical to get enough sample data to calculate the partition points, which can make sure all the partitions' size are similar. Secondly, if the data skew still happens as the performance of some nodes is not as good as others, in Hadoop, it can migrate the tasks from the lower-performance nodes to higher-performance idle nodes.

7.7 High-Availability (HA)

In computing, the term availability is used to describe the period of time when a service is available, as well as the time required by a system to respond to a request made by a user. High availability is a quality of a system or component that assures a high level of operational performance for a given period of time. One of the goals of high availability is to eliminate single points of failure. Typically, High-availability improve the availability of the cluster by having redundant nodes, which are then used to provide service when system components fail.

There are commercial implementations of High-Availability clusters for many operating systems. The Linux-HA project is one commonly used free software HA package for the Linux operating system [15].

A good example of High-availability computing cluster is Apache Hadoop. Hadoop provides High-availability in HDFS system. The HDFS NameNode High Availability feature enables you to run redundant NameNodes in the same cluster in an Active/Passive configuration with a hot standby. This eliminates the NameNodes as a potential single point of failure (SPOF) in an HDFS cluster. Formerly, if a cluster had a single NameNode, and that machine or process became unavailable, the entire cluster would be unavailable until the NameNode was either restarted or started on a separate machine. This situation impacted the total availability of the HDFS cluster in two major ways:

- In the case of an unplanned event such as a machine crash, the cluster would be unavailable until an operator restarted the NameNode.
- Planned maintenance events such as software or hardware upgrades on the NameNode machine would result in periods of cluster downtime.

HDFS NameNode HA avoids this by facilitating either a fast failover to the new NameNode during machine crash, or a graceful administrator-initiated failover during planned maintenance [16].

7.8   SLA

SLA (Service Level Agreement) is an agreement between consumer and service, which warrants generic service functionality. An SLA can be flexible and altered according to the different kinds of services as per the requirement. The purpose of an SLA is to offer evidence that keeps track records of performance, availability and billing. Because of its adaptable quality, a vendor can regularly update its services like technology, storage, capability and infrastructure. By means of negotiation, the consumer and the service provider will agree upon common policies in SLA. The termination phase in SLA delivers the end date of a service and offers the final service bill of utilized resources. It is an easy way to form a treaty between both parties [9].

To guarantee the service quality, some service providers allow customers to submit the SLA together with a job or workload. SLA is used to check whether the service provider can accommodate the job to meet the SLA. If it can, then the service provider executes the job using the SLA. If not, the consumer is asked to negotiate with the service provider to come up with an SLA that both parties could agree upon.

SLA can improve customers' satisfaction. For example, if a user submits a job and expects the job to be finished in a certain time, like 1 h, but due to high usage of the cluster, the job is not completed within 1 h, so the customer is not satisfied with the service. In such case, if there is a SLA to identify the job's requirement and the available resource in the service provider, then the service provider can adopt some alternative methods to meet customer's need, such as adjusting the priority of the job or adding more hardware resources.

In summary, performance, Interoperability, fault-tolerance, Security, Manageability, Load-Balance, High-Availability and SLA are the key Quality of Service aspects those contribute to the success of a well designed Big Data Analytics system.

8 Conclusion

In summary, the concepts discussed in this chapter are fundamentals to Distributed Computing area. The Big Data Technologies implements these concepts and address the quality of services, like performance, fault tolerance, high availability, load balancing, and others while used to address the needs of real life applications and use cases. So understanding these fundamental concepts of Distributed Computing is very important for appropriate use of them in industries specific Big Data Analytics systems. Also using the right trade of across various quality of services is of paramount importance while applying these concepts in the context of specific Big Data Analytics use cases.

References

1. https://www.britannica.com/technology/multiprocessing
2. https://en.wikipedia.org/wiki/Multithreading_%28computer_architecture%29
3. http://www.w3ii.com/en-US/operating_system/os_multi_threading.html
4. https://en.wikipedia.org/wiki/MIMD
5. http://essaymonster.net/science/69515-study-on-mimd-and-shared-memory-architectures.html
6. https://www.techopedia.com/7/31151/technology-trends/what-is-the-difference-between-scale-out-versus-scale-up-architecture-applicat
7. MEN170: SYSTEMS MODELLING AND SIMULATION. QUT, SCHOOL OF MECHANICAL, MANUFACTURING & MEDICAL ENGINEERING
8. Queueing systems and networks. Models and applications. B. FILIPOWICZ and J. KWIECIEŃ
9. https://www.researchgate.net/publication/273575710_Adaptive_Scheduling_in_the_Cloud_-_SLA_for_Hadoop_Job_Scheduling
10. http://robertgreiner.com/2014/08/cap-theorem-revisited/
11. https://mytechnetknowhows.wordpress.com/2016/05/31/cap-theorem-consistency-availability-and-partition-tolerance/
12. https://en.wikipedia.org/wiki/CAP_theorem
13. https://codahale.com/you-cant-sacrifice-partition-tolerance/
14. https://www.techopedia.com/definition/6581/computer-cluster
15. https://en.wikipedia.org/wiki/Computer_cluster
16. https://docs.hortonworks.com/HDPDocuments/HDP2/HDP-2.3.4/bk_hadoop-ha/content/ch_HA-NameNode.html

17. https://www.techopedia.com/definition/631/interoperability
18. https://www.qubole.com/blog/big-data/hadoop-security-issues/
19. http://blingtechs.blogspot.com/2016/02/cap-theorem.html
20. http://mesos.apache.org
21. https://hadoop.apache.org/docs/r2.7.2/hadoop-yarn/hadoop-yarn-site/YARN.html
22. https://zeppelin.apache.org/
23. https://www.cloudamqp.com/blog/2016-09-13-asynchronous-communication-with-rabbitmq.html

Distributed Computing Patterns Useful in Big Data Analytics

Julio César Santos dos Anjos, Cláudio Fernando Resin Geyer, and Jorge Luis Victória Barbosa

1 Introduction

Data-intensive applications like petroleum extraction simulations, weather forecasting, natural disaster prediction, bio-medical and others research have to process an increasing amount of data. In view of this, Big Data applications lead to the need to find new solutions to the problem of how this should be carried out, related to the point of view of dimensions such as Volume, Velocity, Variety, Value and Veracity [1]. This is not an easy task, *Volume* depends on a hardware infrastructure to achieve scalability and *Value* depends on how much Big Data must be creatively and effectively exploited to improve efficiency and the quality needed to assign *Veracity* to information. Variety of data typically originate from different sources, such as historical information, pictures, sensor information, satellite data and other structured or unstructured sources. MapReduce (MR) [2] is a programming framework proposed by Google that is currently adopted by many large companies, and has been employed as a successful solution for data processing and analysis. Hadoop [3] is the most popular open-source implementation of MR.

Since there is a wide range of data sources, the collected datasets have different noise levels, redundancy and consistency [4]. New platforms for Big Data like Cloud Computing (Cloud) have increasingly been used as a platform for business applications and data processing [5]. Cloud providers offer Virtual Machines (VMs), storage, communication and queue services to customers in a pay-as-you-go

J.C.S. dos Anjos (✉) • C.F.R. Geyer
UFRGS, Federal University of Rio Grande do Sul, Institute of Informatics – PPGC,
Porto Alegre, Brazil
e-mail: jcsanjos@inf.ufrgs.br; geyer@inf.ufrgs.br

J.L.V. Barbosa
UNISINOS, University of Vale do Rio dos Sinos, Applied Computing Graduate
Program – PIPCA, São Leopoldo, Brazil
e-mail: jbarbosa@unisinos.br

scheme. Although, Cloud has grown rapidly in recent years, it still suffers from a lack of standardization and the availability of homogeneous management resources [6]. *Private clouds* are used exclusively by a single organization, that keeps careful control of its performance, reliability and security, but might have low scalability for Big Data analytics processing requirements. *Public clouds* have an infrastructure that is based on a specific Service Level Agreement (SLA) which provides services and quality assurance requirements with minimal resources in terms of processing, storage and bandwidth. The Cloud Service Provider (CSP) manages its own physical resources, and only provides an abstraction layer for the user. This interface might vary depending on the provider, but maintains properties like elasticity, insulation and flexibility [7]. On the other hand, *Hybrid clouds* are a mix of the previous two systems and enable the *cloud bursting* application deployment model, where the excess of processing from the *Private cloud* is forwarded to the *Public cloud* provider. Cloud providers can negotiate a special agreement as a means of forming a *Cloud federated* system, where providers that operate with low usage, might be able to lease a part of their resources to other federation member to avoid wasting their idle computational resources [6].

For last few decades, finding information in large datasets was only possible through a relational database. The data professionals had to choose the right query to obtain the correct result. However, in the recent past, it was realized that Relational Database cannot be a tool for every type of domain and analytics problem. In Big Data, the queries can include both structured, semi-structured or unstructured data, such as audio, video, web pages, text and so on, and it can originate from multiple data sources. Multimedia, social networks and Internet of Things (IoT) are collecting more and more information, which means that Big Data will have a growing prospect of being able to create value for businesses and consumers [4]. The purpose of Big Data Analytics is to amass a lot of data and find anomalies or patterns in it, so that value and significance can be added. However, it is common to find multiple data in different places, since the cost of data transfers for a single site is prohibitive owing to the limitations of size and bandwidth [8, 9].

In addition to Cloud, several other types of infrastructure are able to support data-intensive applications. Desktop Grids (DGs), for instance, have a large number of users around the world who donate idle computing power to multiple projects [10]. DGs have been applied in several domains such as bio-medicine, weather forecasting, and natural disaster prediction. Merging DG with Cloud into *Hybrid Infrastructures* could provide a more affordable means of data processing. Several initiatives have implemented Big Data with Hadoop as a MR framework, for instance [11–13]. However, although MR has been designed to exploit the capabilities of commodity hardware, its use in a *Hybrid Infrastructure* is a complex task because of the resource heterogeneity and high churn rate of desktops. This is usual for DGs but uncommon for Clouds. *Hybrid Infrastructures* like these are environments which have geographically distributed resources [9] in heterogeneous platforms with a mixing of Cloud, Grids and DG.

Frameworks and engines to Big Data follow known primitives in computer science such as mechanisms to message synchronization, data distribution, task management and other. The message exchange is the basis of distributed systems, and primitives, like *send* and *receive*, are found built-in on the programming languages in the different frameworks. However, these primitives are only a part of these systems used for the data intensive processing which most of the time, remain hidden to users and programmers. This Chapter introduces some of these primitives and their possible implementations.

The Chapter is organized as follows. Sections 2 and 3 are about primitives for Distributed Computing. Section 2 shows an overview about the main primitives for concurrent programming. Section 3 discusses protocols and interfaces for message exchange. Section 4 presents the data distribution in Big Data over geographically distributed data environments. Section 5 approaches possible implementation problems in distributed Big Data environments. Finally, Sect. 6 presents conclusions.

2 Primitives for Concurrent Programming

The primitives and patterns of Big Data Programming models can be classified into three main areas: concurrent expression and management, synchronization of concurrent tasks and communication between distributed tasks. In this section we'll delve into them in detail.

2.1 Concurrency Expression

The primitive-fork concept allows the creation of a new process within a program.

Other primitives related to this concept enable the execution of another program (executable code), creation and execution of the process on a remote (distributed) computer, and waiting for the termination of a child process. At first, because the process concept does not allow the sharing of variables (data) between two processes, special libraries were created for the declaration of shared variables between processes. In a second moment, the multi-threaded programming model emerged which made concurrent programming much simpler and more efficient, in particular by the ease of native variable sharing. This model was implemented in several instances, highlighting the POSIX threads library, later the Sun Java threads, and then the Microsoft C# threads. When a process is created in the local memory of a machine, a thread is automatically launched as a parent thread. Figure 1 shows a parent thread (*thread A*) which can create one or more child threads (*thread a'*) to the data sharing and a parent thread created by a process (*Proc 2*).

Fig. 1 Processes and
Threads in a local memory

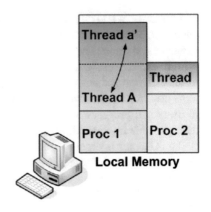

2.2 *Synchronization*

The concurrent programming model with shared variables introduced the synchronization problem. With the increasing popularity of this model, the search for better mechanisms for synchronization has been intensified. There are two major problems of synchronization: the effects of concurrent access of writing to a shared variable and the dependence of one task on results produced by another task. Several authors such as Dijkstra, Hoare, and others have proposed different solutions such as mutex, condition variables, semaphores, and monitors, which have been implemented in various libraries such as Posix threads, Java and C#. For some more specific patterns of concurrency between tasks, other synchronization mechanisms such as barriers and latches have emerged.

For the implementations to be efficient, some evolution in the processors (hardware) was necessary. A good example was the introduction of *TestAndSet* instruction that allows reading and writing in a simple variable (boolean, integer). A great reference for these concurrent programming concepts and their instances is the book by Gregory Andrews [15]. More recently, with the advent of multicore processors and GPUs, there have been some interesting variations to solve the problems of synchronization in both hardware and software. A good example is the concept of transactional memory.

It is important to note that the development of distributed applications requires other primitives of resources and services beyond those presented above, with a particular focus on programming. A classic example is the concept of distributed file systems and their realizations such as NFS, another solution adopted by the Sun company. Also in programming terms, the popularization of systems and applications in local and wide networks, that is, a set of distributed and independent computers required the development of the message-based programming concept, which will be presented in the next section. However, most of the primitives mentioned above for concurrent programmings, such as for thread creation and management, and for synchronization of shared variables, do not have satisfactory variants for the context of distributed systems.

3   Communication Protocols and Message Exchange

The distributed messaging-based programming model allows two or more processes, or programs, running on separate computers, without access to the concept of shared memory, to exchange information. Using the specific model of the *send/ receive* primitives, a *sender* process sends, through the *send* primitive, to an identified *receiver* process a data that it has in its local memory. The *receiver* process receives a copy of the data through the *receive* primitive and stores it in its local memory. Usually the *receiver* process does not need to identify the *sender* process. This principle is a basic, simple, and abstract model, as exemplified in Fig. 2.

There are numerous variations derived from this basic model considering aspects such as the synchronization that may occur between processes during the execution of *send/receive* primitives. In addition to model variations, the study of message exchange concepts is still more complex if one considers the numerous instantiations (implementations). They can be differentiated for example by the different communication network protocols that can be used in the execution of the *send/ receive* primitives, such as TCP or UDP protocols.

3.1   Synchronous Communication

In the *synchronous communication* the sender and receiver are synchronized in each message. The sender is blocked until the receive primitive is executed by the receiver process, delivering a copy of the data to the application. The sender process can execute a second (next) *send* only after the receiver has executed the receive primitive for the precedent message. Usually the receiving process is blocked by the receive primitive until a message has been delivered by the system to the receiving process. However, the sender and receiver devices must have a temporization mechanism when multiple hosts are sending messages simultaneously, such as an atomic clock, to avoid the arrival of out-of-order messages. Also, another possible problem is when the sender remains blocked all the transmission time due to an inefficient synchronization problem. Thus, the system could have synchronization problems mainly on the Internet environment which is very unstable.

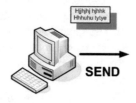
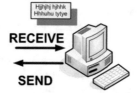

Fig. 2   Send and Receive Primitives

Also, the *synchronous communication* is uncommon in the most of the distributed system and the synchronous model it is inefficient in many of the distributed systems [16].

3.2 Asynchronous Communication

In the *asynchronous communication*, the *send* process is unlocked and can continue its execution (following the *send* command) soon after the send execution is passed to the service that implements the complete message exchange functionality. Usually, this service starts on the sending computer (operating system and network services) when a complete copy of the data is made by the service, allowing the *send* process to update the data sent in its local memory. The most implementations of the basic model adopt this semantic.

An important question is how processes identify the partner process in the *send* and *receive* primitives. For example, each process may have a unique identifier (ID) in the context of the system or application. This ID is generated dynamically at the time that the process is created and can be assigned by the programmer. Another option is to use a computer identifier, where the process is running, associated with a communication port allocated to that process. Other issues that distinguish the different implementations and that may consequently affect the techniques of use by the application are:

- What is the maximum size of each message?
- What types of data can be transmitted in a message?
- What control over any data transmission errors, such as message loss or duplication, and content change?
- Can sender and receiver processes be implemented in different languages and run on distinct platforms (*e.g.* operating systems)?

3.3 Pseudo-Synchronous Communication

A variant of the two previous models was created in the context of some *message oriented middleware* (MOM). One of the advantages of *synchronous communication* is that the sending process, shortly after the execution of the *send* primitive, knows that its message has been received by the receiver. The same does not occur in *asynchronous communication*. In the *pseudo-synchronous communication* variant, the sending process is not blocked at each *send*. But the receiving processor sends asynchronously to the sender a message confirming the receipt of each message sent by the sender. This confirmation message can have multiple meanings, for example, the message was only received or the message was received and processed. It is important to note that in the literature it is possible to find relevant variants for the concept of *pseudo-synchronous communication*. In Distributed System theory, some authors define three systems models: *synchronous, asynchronous* and *partially synchronous communication*. The concept of communication in *partially synchronous* systems just expresses that the sender process is not blocked (as in the *asynchronous* model), but there is a time limit for communication to be completed, while in *asynchronous* this time is theoretically "infinite."

3.4 Client/Server Paradigm

In the context of the *Client/Server* model, for the distributed systems architecture, a new paradigm of communication called Remote Procedure Call (RPC) has emerged. In RPC, a client program makes a call to a procedure located on a remote server. The client is usually blocked until the procedure is completely executed. Communication is optional, through procedural arguments, and obviously can be performed in both directions. The client must know the server's remote procedure interface, that is, a procedure name and types of input and output arguments. Initially an implementation was developed over the TCP protocol (which guarantees greater reliability than UDP) and applied in the development of several important distributed systems, such as NFS. Later on, variations of the basic RPC model, better adapted to new distributed system needs, emerged, such as the Web Services concept, more appropriate than the classic RPC for web-program communication.

The data serialization process in the distributed file system of Hadoop, a framework extensively used in Big Data, produces byte streams for transmission over a network or for writing to persistent storage. The inter-process communication between nodes in the system is implemented under RPC to execute these tasks [3].

Finally, it is necessary to at least mention some models created for more specific purposes such as replication (data or servers), fault tolerance, or in the context of more recent computing paradigms such as Mobility, P2P, Cloud and IoT. Examples are libraries for group communication, tuple spaces, *publish/subscribe*, location, event processing and others. The following subsection discusses platforms that support communication in Big Data environments.

3.5 Communication Deployment in Big Data

The Apache Flink, previously called Stratosphere, is a data analytics framework that follows the *Lambda Architecture* and enables the extraction, analysis and integration of heterogeneous datasets [17]. It has two APIs, one for *DataSet* and other for *DataStream* respectively deployed on process batch and stream applications which constitute a hybrid programming environment. The core is a dataflow in a distributed streaming that does not store data but converts it into optimized binary formats, after its reading. It is extensible for traditional data warehousing queries such as textual data queries and information integration in a Table API library. The implementation supports iterative programs that allow an analysis of execution graph and statistical applications inside the data processing engine. The architecture includes different deploying modes, such as local (in a single JVM), cluster (standalone and Hadoop-YARN environment) and Cloud computing (EC2 and GCE). Flink supports Java, Scala and Python programming languages [18]. Its implementation has a publish/subscribe API to connect with Apache Kafka [19], which is used as a data synchronization mechanism in heterogeneous environments.

MapReduce is a programming framework that abstracts the complexity of parallel applications. It is a batch processing system that partitions and scatters datasets across hundreds or thousands of machines, bringing the computation and data as close to each other as possible [3]. The *Map* and *Reduce* phases are handled by the programmer, whereas the *Shuffle* phase is created while the task is being carried out. The input data is split into smaller pieces called chunks. The data is serialized and distributed across machines that form the Distributed File System (DFS). The Hadoop implementation which follows the *MapReduce* model uses a synchronization mechanism through heartbeat between Master and Workers.

The master handles the task scheduling according to the data locality. Therefore the tasks are running locally in the first phase. When running an application, the master assigns tasks to workers and monitors the progress of each task. The machine that is assigned a *Map* task, executes a *Map* function and emits *key/value* pairs as intermediate results that are temporarily stored in the workers' disks. The execution model creates a computational *barrier*, which allows the tasks to be synchronized between the producers and consumers. A *Reduce* task does not start its processing until all the *Map* tasks have been completed. This works as a data synchronization mechanism. A hash function is applied to the intermediate data to determine which key will carry out a *Reduce* task. The group of selected keys forms a partition. Each partition is transferred to a single machine during the *Shuffle* phase, to execute the next phase. The serialization function is an RPC client/server implementation which uses the primitives *send/receive* to do these transfers.

Spark [20] is a Big Data implementation widely popular to real-time applications. The main abstraction in Spark is the RDD (called resilient distributed datasets) [21], a storage abstraction that avoids replication by using lineage for fault recovery, *i.e.*, the events are grouped into micro-batches. The RDD is kept in memory not as a distributed shared memory abstraction, but as objects partitioned across a set of machines that can be rebuilt if a partition is lost. RPC is the main communication paradigm used to access these objects. The programmer produces operations with "map, filter and join" and enables interactive data mining. Although RDD are best suited for batch applications that apply the same operation to all elements of a data set; they are less suitable for applications that make asynchronous fine-grained updates to shared state.

Spark maintains the data synchronization and manages resource allocation in Big Data real-time applications. However, more complex environments use the Zookeeper [22] system to provide leader election and some state storage with multiple masters. Zookeeper is a service that maintains the configuration information and provides the synchronization to distributed applications. This system enables the recovery process occurs between 1 and 2 min in failure cases. Thus, the complex implementations of race conditions in Big Data applications on distributed systems are hidden from users and programmers.

4 Data Distribution in Big Data on Distributed Environments

Big Data applications can be implemented in several ways. Scattered data can be found in DNA research studies, where researchers need to investigate different databases, such as those in protein structure analysis. These applications seek a genetic mapping that requires a pre-existing reference genome to be employed for the read alignment of a gene [23]. Thus, the data processing is characterized by its ability to compare input data with different databases. This processing consists of several phases of search-merge-reduce, where the data are given an incremental update [24]. Some researchers like Jayalath [8], Tudoran *et al.* [25], Krish [26] and Ji [27] have put forward Hadoop implementation based on a geo-distributed dataset in multiple data centers. The authors state that, for instance, it is possible to have multiple execution paths for carrying out a MapReduce job in this scenario, and the performance can carry out a great deal.

Figure 3 illustrates the scenario where dispersal data is used. Each locality is connected through slow links, where data transfers may not have a negligible cost. The data is scattered in the clusters. All the intermediate results must be combined to produce a single return for a Big Data analysis. These problems can be overcome by means of a hybrid infrastructure if there is a file system that supports the incremental

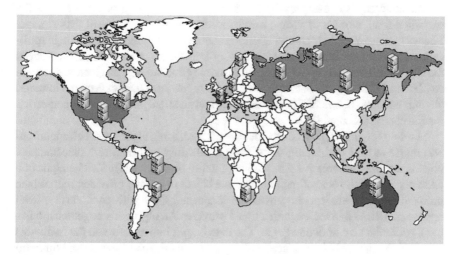

Fig. 3 Geographically distributed data

updates and highly concurrent data sharing. A possible solution involves integrating a distributed file system like the Hadoop Distributed File System (HDFS) with the use of a Cloud environment; otherwise, DGs are a large-scale infrastructure with specific characteristics in terms of volatility, reliability, connectivity, security and storage space. Both architectures are suitable for large-scale parallel processing. Finally, more complex combinations can be envisaged for platforms resulting from the use of multiple Clouds through an extension to a DG [28, 29]. However, these structures must use several synchronization primitives such as those discussed in Sect. 2.

Different Cloud infrastructures have their own configuration parameters, and the availability and performance of offered resources can change dynamically due to several factors, including the degree of over-commitment that a provider employs. In this context, solutions are needed for the automatic configuration of complex cloud services. The Cloud infrastructure comprising heterogeneous hardware environments may need the specifications of configuration parameters at several levels such as the operating systems, service containers and network capabilities [30]. The users who need to execute applications may not know how to map their requirements to the available resources. The lack of knowledge about the cloud provider infrastructure will lead either to overestimating or underestimating the required capacity; both are equally bad and can lead to a waste of resources.

Organizations are increasingly relying on an infrastructure from multiple providers as a means of increasing fault tolerance and avoiding provider lock-in. A *Multi-Cloud* infrastructure contains various configuration choices and can change its requirements and workloads dynamically at the time of execution. Given this, solutions are needed for the automatic configuration of complex cloud services at different abstraction levels. In this context, multiple cloud infrastructures, like clouds in heterogeneous environments, require different configuration levels and processing synchronization such as the operating system, service containers, and configuration capabilities [30].

The allocation of resources from CSPs to users is carried out in terms of the execution time, number of virtual machines, data transfer and size of data storage. The users must map their computational resource needs before running their applications. This means that, if there is a lack of knowledge on the part of users about the CSP infrastructure or a real need for resource allocations, it can lead to an incorrect lease of CSP resources for the users and a higher cost than expected. However, an optimal allocation is difficult to achieve, and so strategies to obtain an approximation can be accepted [31].

Mansouri *et al.* [32] propose the deploy of a brokering algorithm. This algorithm was employed for optimizing the storage availability and finding a placement of objects that was suitable for the required Quality of Service (QoS). The algorithm takes account of the cost of maintaining one object in a cloud provider, reduces the probability of failure and improves the associated QoS with each service-level agreement (SLA) contracted with a cloud provider. An object is a target data, without a particular size or defined type. The data is split into *chunks* and the main goal is to find the optimal *chunk* placement depending on the user's needs and financial means.

A large number of transfers of objects from one cloud storage provider to another takes up time and is often impossible during the execution time. An expected availability represents M objects in each data center, and this determines the expected failure of the object in each data center. Mansouri *et al.* evaluate two parameters to each cloud provider, the failure probability and the cost per object. The objects are replicated in multiple sites in accordance with these metrics. However, the total size can achieve up to several exabytes, which can require a lot of time for these transfers.

The message synchronization in some cases is difficult to do when several systems are working together with different topologies under the Internet. One reason is that can have different time synchronization requirements on several applications simultaneously. In this case, an orchestrator is required to maintain this control. SALSA [30] is a framework for the orchestrated configuration of cloud services through multiple CSPs. This framework provides a model for application configurations and the deployment of different kinds of services. The information about the configuration supports each level of cloud service such as application levels, deployment relationships at multiples software stacks and the link between service units and configuration capabilities. The configuration capabilities are obtained from registered services (cloud services and specifications of topology services) or user specifications. SALSA has a *service unit orchestrator* for multiple configuration services for each configuration task group. Its purpose is to control the application deployments, movement of virtual instances among different cloud providers and the deployment of an environment like VM, library loads and support for multiple stack deployments of cloud.

The creation of VM is a separate process from other software levels. The configuration capabilities can be obtained via a registry service or from user specifications, to determine the relationship between the service units. A service orchestrator is generated for each service allowing it to handle the tasks. *Meta Information* contains abstract nodes with generic types of service that implement the virtual nodes. SALSA adopts an approach where each *service unit orchestrator* runs independently and interacts with a cloud service orchestrator. Although the framework enables heterogeneous configurations, there is not a mechanism to evaluate the performance or the workloads used to adapt the load-balance in Cloud. The SALSA architecture, designed by [30], is shown in Fig. 4.

The framework has a central configuration service that orchestrates the setup operation through the local configuration layer in the VM. The information service keeps a good deal of information about the Cloud infrastructure which is handled by a configuration generator. The topology orchestration layer creates a dependency graph and sets a configuration plan for a Cloud configuration system and the VMs are managed in this way. The monitor layer keeps the status of both VMs and the Cloud elasticity but it is necessary for the services to be already working before new service instances can be distributed.

HyMR [33] is a framework for enabling an autonomic *cloud burst* for clusters of virtual machines that execute MapReduce jobs over *Multi-Cloud*. The authors implemented a Hybrid Infrastructure as a Service (HyIaaS) for the VM instance

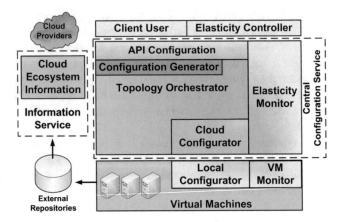

Fig. 4 The Salsa architecture (Adapted from [30])

(partitions management) in *Multi-Cloud*. HyIaaS implements an OpenStack[1] extension. This partitioning is transparent to the users, since it allows them to have access to all the VMs in the same way, regardless of their physical allocation. HyIaaS receives the deadline specifications of the users that are stored in a user-policy for managing VM migrations. An external CSP will be responsible for receiving and launching the VMs across their *Cloud Controller* module.

The *Logical Node* monitors and analyses critical events from a physical machine and the *Logical Cloud* makes spawning/migration decisions based on *Logical Node* information. Figure 5 shows the HyMR architecture, where a HyIaaS orchestrates the application executions. The HyMR runs on the *Cloud Controller* and maintains data consistency in a part of HDFS. However, VM migrations have a poor performance when carrying out data copying operations from the HDFS.

These approaches show the need for a more fine-grained system of task management and data distribution across Big Data applications. Businesses and governments arrange their data in distributed cloud platforms for different reasons, such as, the need to maintain the proximity of resources; data storage with organizations that share common goals; and a desire to keep data replicas across regions for redundancy purposes. This data information must be analyzed on a global scale.

One possible way to do this is to merge all the data in a single data center, and another is to use a *Multiple Clouds* infrastructure to execute individual instances of *MapReduce* across each dataset separately and then aggregate the results. The study of [8] suggests that this could be done by running jobs in a geo-distributed operation. The authors introduce the G-MR, a Hadoop implementation based on a geo-distributed dataset across multiple data centers. They state that, for instance, it is possible to have multiple execution paths for carrying out a *MapReduce* job in this scenario, although the performance may vary considerably for each path. Another problem is that popular *MapReduce* open sources, like Hadoop, do not support this

---

[1] https://www.openstack.org/.

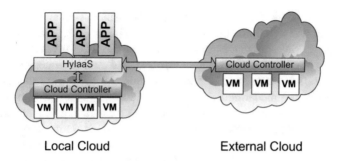

Fig. 5 The HyMR architecture (Adapted from [33])

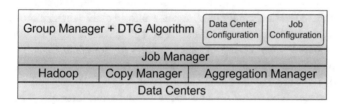

Fig. 6 The G-MR architecture (Adapted from [8])

feature. In addition, most CSPs do not usually provide a bandwidth guarantee for large-scale data transfers in execution time [34].

The G-MR has an algorithm called a *Data Transformation Graph* (DTG) which determines an execution path for performing a job sequence for *MapReduce*. The problem is how to decide which stage should derive partitions that must be moved and how to reduce costs by finding the best performance for *MapReduce* applications. Figure 6 shows the architecture of G-MR, adapted from [8]. The architecture consists of the following modules: a *Group Manager*, *Job Manager*, *Copy Manager* and *Aggregation*. The *Group Manager* optimizes the execution path and may instruct the *Job Manager* to copy data for a remote data center or aggregate multiple sub-datasets. The *Job Manager* performs the jobs over Hadoop which is deployed in each *n* data center. The *Copy Manager* is responsible for executing the data copy from one data center to another. However, the total number of nodes in a single job graph is $O(p^n)$ and can become huge when the number of *p* partitions grows. The *Aggregation* manager maintains the integrity of the results. The model shows that this architecture is feasible from the standpoint of data distribution and the integration of results.

The approach *Write Once Read Many* (WORM) is an accepted assumption for data access in *Big Data* applications like *MapReduce*. The handiest manner for *Big Data* processing across several data centers is to use a data replication mechanism among different CSPs. However, the variability in the high-performance required for cloud operations leads to bottlenecks [35, 36]. Thus, the best strategy is to reduce data transfers.

Tudoran *et al.* [25] argue that there are two methods for modeling complex infrastructures. The *analytical models* use low-level details with workloads and are characterized by their ability to predict the performance. This means the wealth of detail is what will determine the best modeling. The *sampling method* is an active approach which does not require any previous knowledge of the infrastructure. The information about network bandwidth, topology and routing strategies is not available to the users in public clouds. Because of this, the authors introduce a sample-based category for modeling that monitors the environment with agents, called GEO-DMS. The agents carry out the monitoring for data transfers and geographically-distributed data management that is transferred across multiple clouds. The model registers the correlation between performance (execution time) and cost effectiveness (finance), and imposes budgetary constraints in the interests of safety.

The agents are implemented through VMs in each CSP where the applications are running. The *decision manager* is concerned with how the transfer paths are established between the source and destination. One way to achieve this is directly from the node to the data center or by using multiple paths across intermediate data centers. The data transfers are intra-site data replications that result from the presence of dedicated links among the data centers of the same CSP. The scientific applications interact with an API to provide data transfers over a WAN. A *monitor agent* carries out environmental monitoring and makes the measurements for the decision manager. The measurements include bandwidth throughput between data centers, and the CPU load, I/O speed and memory status of the VM nodes. The *decision manager* updates the weights of the paths periodically with the aid of these measurements.

5   Implementation Problems

This section has been organized to provide the first vision about some paradigms behind of the Big Data implementations and discuss possible problems that the developers must avoid. As previously discussed, in a distributed environment we can have several implementation issues. The biggest problem of the resource sharing is related to the need to avoid the race condition between the systems and the loss of synchronization on message exchange problems. With basis on these issues, the next sections evaluate possible implementation problems in Big Data.

5.1   Race Condition Problems

The reduction of performance in virtualization systems has led companies like Google and Facebook to use physical machines directly. The first reason for performance decrease is the *lock* in the disk access due to race conditions. The semantic follows the *send/receive* paradigm in an *asynchronous communication* channel. The operating system of the virtualization manager provides the access control to the internal bus for one single machine on each time where to do write on disk. If the data size is larger than the queue size, then several I/O interruptions will be necessary. As more than one virtual system can access the same resource, the wait time to use the disk queue increases and produces a high wait time to each virtualized application. Thus, this behavior will produce a low performance. This phenomenon also occurs when the cloud providers use storage under fibre channel networks.

Hybrid systems have been described in some scientific papers as a mixture of public and private clouds. At the same time, this factor refers to the degree of availability in the resource policies. Factors regarding the deployment of native and virtualized clusters are evaluated in the work of Sharma *et al.* [37]. In their analysis, the authors argue that generic benchmarks show an overhead of 5% and 15% for computation and I/O workloads respectively, when confronted with a non-virtualized system. In addition, the level of overhead may vary depending on the workload, availability of resources and programming of interactive jobs. This I/O competition is related to the bandwidth of the disk bus buffers in the hardware environment which generates a race condition problem, called of disk contention problem.

There is an increasing need for a repeated analysis from Big Data in the Cloud with streaming characteristics. Streaming and data-intensive applications are often not the best profile for Cloud applications [38]. Streaming systems are event-driven and their behavior differs from batch systems like *MapReduce* [39]. The *MapReduce* model lacks efficient support for real-time processing. The traditional system that has been developed to process static databases like Hadoop cannot provide a low latency response in real-time or streaming processing. The main problem in streaming is the *lock* in the receiving mechanism due to the over-information in the input queue during an event where occurs a data burst. Another problem is related with system incapacity to forecast the exact moment of occurring a data burst. Some implementations avoid this problem with the brokers use like the Kafka, which isolates the producers and consumers in a message queue system. This application can produce small overhead in comparison with achieved benefits.

Hadoop Streaming is a utility included in the Hadoop distribution in an attempt to enable streaming in the MapReduce model through two standard Unix interfaces for stream processing, one *stdin* (input) and one *stdout* (output) [40]. These interfaces have two "executables" (one mapper and one reducer) that use a *Combiner* function. The *Combiner* is an implementation that enables a map output to run in memory. *Pipe operations* are created by the Unix system call to build a *half-duplex communication channel* for an external executable file. A *pipe call* is invoked to establish a *communication channel* for each interface with *send/receive* primitives.

When a new *pipe* is created, two file descriptors are obtained. One of them is for reading and the other for writing. An *inode* identifies each *pipe* in the local file system to temporarily store data. The *communication channel* may generate system overhead producing a *lock* and some critical fields of *pipe* can spawn race conditions between the read/write operations. A manner to avoid this issue is to increase the size of the buffer greater than the queue length in the system memory.

5.2   Message Exchange

The *MapReduce* model was originally conceived for large homogeneous cluster environments. As a result, simplifications were adopted by the model with the aim of optimizing the task distribution. However, these simplifications may entail system degradation in heterogeneous environments. The work of Zaharia *et al.* [41] was the first study to detect these problems. Their study points out that there are concerns over the simplification of the *MapReduce* model since this may lead to an excessive number of speculative tasks. These issues were observed among the executions of different applications in large *clusters* with virtual environments, *e.g.* Amazon *EC2*, owing to the competition for applications that provide access to hardware. The Cloud resources represent the most homogeneous environment possible, although, in fact, the resources have a heterogeneous behavior because they share virtual machines with other users at the same time. To overcome this problem, the authors proposed LATE (*Longest Approximate Time to End*), a new task scheduler. Although LATE does not completely avoid speculative tasks, it considerably reduces performance degradation in heterogeneous environments. The results of an experimental evaluation results show that, compared with the native Hadoop scheduler in speculative mode, LATE achieves a gain ranging from 8.5% to 58%, depending on the application and number of working machines and thus maximizes the message exchanges mechanism and avoids synchronization loss.

In the work of Tang *et al.* [42] the authors implement a layer to create a hybrid distributed file system (*HybridDFS*) under reliable data storage clusters in Cloud and unreliable data storage in desktop grid. Each data chunk in volatile nodes has at least one replica distributed for different volunteer PCs or cluster nodes. The volatile node employs a fault-tolerance mechanism.

The authors developed a Node Priority-Based Fair Scheduler (NPBFS) algorithm. This means that the node with greater computational capacity processes more tasks. The system takes into account both data location and storage capacity, and the length of the task queues is related to the computational capacity of each node. A weight coefficient indicates the node priority for receiving a greater number of tasks. The algorithm works as a FIFO queue, which maintains the task distribution under a *REST Web Server*. The REST Web Server is a structured message exchange service in computer networks following the *Client/Server* paradigm. The management of the data transfers follows a synchronization mechanism based on two thresholds: the synchronization interval time (SIT) and failure timeout time (FTT). An FTP service provides the data movement.

In a *Hybrid Cloud* environment, the data centers are interconnected by means of slow links. The data is moved from the private to public Cloud when a new VM allocation is necessary to improve a task performance. The data locality and data movement are a challenge for accelerating iterative MapReduce in *Hybrid Clouds*. Furthermore, since the extra resources represent an additional cost for data movement, a trade-off between performance gains and benefits must be evaluated. These issues are evaluated in the work by Clement [43] to address iterative MapReduce problems in a Hybrid IaaS Cloud environment. The authors argue that improving the ability to take advantage of data locality in a hybrid Cloud environment is critical. The aim of the strategy is to extend the original fault-tolerance mechanism of HDFS and deploy data replicas from an on-premise VM in a private Cloud to another allocated off-premise VM in a public Cloud as if it was an external rack over the HDFS.

The off-premise VM initializes without data and needs re-balance the initial data blocks with on-premise VM. A heuristic determines a re-balance factor from an I/O intensive benchmark to approximate the application behavior for the duration of the re-balancing. The scheduler waits for the off-premise VM to get a minimal replica number to start the task distribution. This deploying is possible due to an application class called iterative application, which reuses the invariant input data, where the data locality can be leveraged. The solution improves the performance with increase parallel executions in a distributed environment in Cloud off-premise. The *message exchange* paradigm is deployed based on RPC, and the message synchronization is based on the heartbeat mechanism of Hadoop.

The strategies for implementing streaming in Cloud are discussed in the work of Tudoran *et al.* [38]. The authors evaluate applications like the Ocean Observatory Initiative, where sensors send information that is collected by satellites for geo-spatial computations. The study shows how communication in the Cloud can interfere with computation. The approach uses persistent and ephemeral storage. In the first, called *Stream & Compute* (SC), the data is sent directly to VM computation without storage persistence. In the second, called *Copy & Compute* (CC), the data is first saved in an attached storage, so that it can be conveyed to VM computation afterwards. When the sensors produce new data, it is processed against existing features as in a temporal process. This eliminates the need for communications between processes, but adds a huge volume of data that must be streamed for each worker. This mechanism uses a Web server in a *client/server* paradigm.

The *Stream & Compute* provides a better response time, but when carried out on a large scale, there is a network saturation that leads to a need for redistribution across different CSPs. The *Copy & Compute* makes it easier to do repairs, when an unexpected stop occurs in the VM. In contrast, the date is near to the computation. However, a high variability in the remote copy phase causes a variation of around 20% in terms of data transfers per seconds. I/O contention is observed in the *Copy & Compute* approach on disc, when all the workers are trying to access the data at the same time. The classic problem of CPU utilization vs. I/O has a significant impact on the data throughput owing to virtualization. The I/O contention problem is typically one of *race conditions* in hardware.

6 Conclusion

This Chapter has presented examples of primitives and patterns used in distributed systems which are implemented in several Big Data engines. The internal codes hide this implementation under methods in Java, C#, and other programming languages. For instance, the synchronizing method, management of variables, inter-process communications, and other primitives. In function of this, the implementations were analyzed in the deployment context of Big Data. Also, the studies on data distribution in the geographically distributed environments have discussed manners to avoid these problems.

Several implementations use the *Client/Server* model and try to avoid race conditions with synchronization mechanisms previously discussed. The standard primitives like *send/receive* and *lock/mutex* are widely used in programming languages like Java, C# and other. These program languages are the basis for the development in Big Data. For instance, a significant problem in Cloud environment is related with I/O contention. However, this issue can not be avoided but only minimized until this moment.

The Big Data applications can be implemented in different ways such as in geographically distributed environments. In this scenario, the control of synchronization can be so complex for the traditional synchronization methods that it is necessary the use of dedicated tools like Kafka and Zookeeper to provide the activities control like the coordination of machines and message synchronizations. The dataintensive management is hard in heterogeneous environments because of the time synchronization. Due to this, the systems must maintain an external clock, such as the atomic clock mechanism with the NTP protocol under the Internet, to avoid the different timestamp between nodes which can generate several synchronization problems in Big Data applications.

Multimedia, social networks and Internet of Things (IoT) are collecting more and more information, which means that Big Data will have a growing prospect of being able to create value for businesses and consumers. The purpose of Big Data is to amass many data and find anomalies or patterns in it, so that value and significance can be added. The emerging systems are highly heterogeneous environments with variable structures, where resources can be added or removed at any time. Thus, it is necessary to find new ways of processing Big Data which can exploit idle computational resources and allow them to be combined with heterogeneous infrastructures in distributed computing.

References

1. M. D. Assuncao, R. N. Calheiros, S. Bianchi, M. A. Netto, R. Buyya, Big Data computing and clouds: Trends and future directions, Journal of Parallel and Distributed Computing 79–80 (2015) 3–15, special Issue on Scalable Systems for Big Data Management and Analytics. doi:10.1016/j.jpdc.2014.08.003.
2. J. Dean, S. Ghemawat, MapReduce - A Flexible Data Processing Tool, Communications of the ACM 53 (1) (2010) 72–77. doi:10.1145/1629175.1629198.
3. T. White, Hadoop - The Definitive Guide, 3rd Edition, Vol. 1, O'Reilly Media, Inc., California, USA, 2012.
4. M. Chen, S. Mao, Y. Liu, Big Data: A Survey, Mobile Networks and Applications 19 (2) (2014) 171–209. doi:10.1007/s11036-013-0489-0.

5. L. M. Pham, A. Tchana, D. Donsez, V. Zurczak, P.-Y. Gibello, N. de Palma, An adaptable framework to deploy complex applications onto multi-cloud platforms, in: Computing Communication Technologies - Research, Innovation, and Vision for the Future (RIVF), 2015 IEEE RIVF International Conference on, 2015, pp. 169–174. doi:10.1109/RIVF.2015.7049894.
6. A. N. Toosi, R. N. Calheiros, R. Buyya, Interconnected Cloud Computing Environments: Challenges, Taxonomy, and Survey, ACM Comput. Surv. 47 (1) (2014) 7:1–7:47.
7. S. Sakr, A. Liu, D. Batista, M. Alomari, A Survey of Large Scale Data Management Approaches in Cloud Environments, Communications Surveys Tutorials, IEEE 13 (3) (2011) 311–336. doi:10.1109/SURV.2011.032211.00087.
8. C. Jayalath, J. Stephen, P. Eugster, From the Cloud to the Atmosphere: Running MapReduce across Data Centers, Computers, IEEE Transactions on 63 (1) (2014) 74–87. doi:10.1109/TC.2013.121.
9. B. Heintz, A. Chandra, R. K. Sitaraman, J. Weissman, End-to-end Optimization for GeoDistributed MapReduce, Cloud Computing, IEEE Transactions on PP (99) (2014) 1–14. doi:10.1109/TCC.2014.2355225.
10. C. Cerin, G. Fedak (Eds.), Desktop Grid Computing, 1st Edition, Numerical Analysis and Scientific Computing, CRC Press, 2012.
11. H. Lin, X. Ma, J. Archuleta, W.-c. Feng, M. Gardner, Z. Zhang, MOON: MapReduce On Opportunistic eNvironments, in: Proceedings of the 19th ACM International Symposium on High Performance Distributed Computing, HPDC '10, ACM, New York, NY, USA, 2010, pp. 95–106. doi:10.1145/1851476.1851489.
12. F. Costa, L. Silva, M. Dahlin, Volunteer Cloud Computing: MapReduce over the Internet, in: Parallel and Distributed Processing Workshops and Phd Forum (IPDPSW), 2011 IEEE International Symposium on, 2011, pp. 1855–1862. doi:10.1109/IPDPS.2011.345.
13. L. Lu, H. Jin, X. Shi, G. Fedak, Assessing MapReduce for Internet Computing: A Comparison of Hadoop and BitDew-MapReduce, in: Proceedings of the 2012 ACM/IEEE 13th Int. Conference on Grid Computing, GRID '12, IEEE Computer Society, Washington, DC, USA, 2012, pp. 76–84. doi:10.1109/Grid.2012.31.
14. W. R. Stevens, S. A. Rago, Advanced Programming in the UNIX Environment, 3rd Edition, Addison-Wesley Professional, 2013.
15. G. R. Andrews, Concurrent Programming: Principles and Practice, Benjamin/Cummings Publishing Company, 1991.
16. N. A. Lynch, Distributed Algorithms, The Morgan Kaufmann Series in Data Management System Series, Morgan Kaufmann Publishers, 1997.
17. A. Alexandrov, R. Bergmann, S. Ewen, J. Freytag, F. Hueske, A. Heise, O. Kao, M. Leich, U. Leser, V. Markl, F. Naumann, M. Peters, A. Rheinlander, M. J. Sax, S. Schelter, M. Hoger, K. Tzoumas, D. Warneke, The Stratosphere platform for big data analytics, VLBD Journal 23 (6) (2014) 939–964. doi:10.1007/s00778-014-0357-y.
18. A. Chauhan, T. Dunning, A. Gates, O. O'Malley, S. Owen, H. Saputra, Apache Flink (2015). URL https://flink.apache.org
19. T. Zhang, Reliable Event Messaging in Big Data Enterprises: Looking for the Balance Between Producers and Consumers, in: Proceedings of the 9th ACM International Conference on Distributed Event-Based Systems, DEBS '15, ACM, New York, NY, USA, 2015, pp. 226–233.
20. M. Zaharia, M. Chowdhury, M. J. Franklin, S. Shenker, I. Stoica, Spark: Cluster Computing with Working Sets, in: 2010 USENIX Federated Conferences Week, 2nd - Workshop on Hot Topics in Cloud Computing, 2010, pp. 1–8.
21. M. Zaharia, M. Chowdhury, T. Das, A. Dave, J. Ma, M. McCauley, M. J. Franklin, S. Shenker, I. Stoica, Resilient Distributed Datasets: A Fault-tolerant Abstraction for In-memory Cluster Computing, in: Proceedings of the 9th USENIX Conference on Networked Systems Design and Implementation, NSDI'12, USENIX Association, Berkeley, CA, USA, 2012, pp. 2–14. URL http://dl.acm.org/citation.cfm?id=2228298.2228301
22. A. S. Foundation, Apache Zookeeper (Jul. 2016). URL https://zookeeper.apache.org
23. Q. Zou, X.-B. Li, W.-R. Jiang, Z.-Y. Lin, G.-L. Li, K. Chen, Survey of MapReduce frame operation in bioinformatics, Journal Briefings in Bioinformatics 15 (4) (2014) 637–647. doi:10.1093/bib/bbs088.

24. A. McKenna, M. Hanna, E. Banks, A. Sivachenko, K. Cibulskis, A. Kernytsky, K. Garimella, D. Altshuler, S. Gabriel, M. Daly, M. A. DePristo, The Genome Analysis Toolkit: A MapReduce framework for analyzing next-generation DNA sequencing data, Genome Research 20 (9) (2010) 1297–1303. doi:10.1101/gr.107524.110.

25. R. Tudoran, A. Costan, R. Wang, L. Bouge, G. Antoniu, Bridging Data in the Clouds: An Environment-Aware System for Geographically Distributed Data Transfers, in: Cluster, Cloud and Grid Computing (CCGrid), 2014 14th IEEE/ACM International Symposium on, Chicago, IL, USA, 2014, pp. 92–101. doi:10.1109/CCGrid.2014.86.

26. K. Krish, A. Anwar, A. R. Butt, HATS: A Heterogeneity-Aware Tiered Storage for Hadoop, in: Cluster, Cloud and Grid Computing (CCGrid), 2014 14th IEEE/ACM International Symposium on, Chicago, IL, USA, 2014, pp. 502–511.

27. S. Ji, B. Li, Wide area analytics for geographically distributed datacenters, Tsinghua Science and Technology 21 (2) (2016) 125–135. doi:10.1109/TST.2016.7442496.

28. G. Antoniu, J. Bigot, C. Blanchet, L. Bouge, F. Briant, F. Cappello, A. Costan, F. Desprez, G. Fedak, S. Gault, K. Keahey, B. Nicolae, C. Perez, A. Simonet, F. Suter, B. Tang, R. Terreux, Scalable Data Management for Map-Reduce-based Data-Intensive Applications: A View for Cloud and Hybrid Infrastructures, Int. Journal of Cloud Computing 2 (2013) 150–170.

29. J. C. S. Anjos, G. Fedak, C. F. R. Geyer, BIGhybrid: a simulator for MapReduce applications in hybrid distributed infrastructures validated with the Grid5000 experimental platform, Concurrency and Computation: Practice and Experience 28 (8) (2016) 2416–2439. doi:10.1002/cpe.3665.

30. D.-H. Le, H.-L. Truong, G. Copil, S. Nastic, S. Dustdar, SALSA: A Framework for Dynamic Configuration of Cloud Services, in: Cloud Computing Technology and Science (CloudCom), 2014 IEEE 6th International Conference on, 2014, pp. 146–153. doi:10.1109/CloudCom.2014.99.

31. L. Mashayekhy, M. Nejad, D. Grosu, A PTAS Mechanism for Provisioning and Allocation of Heterogeneous Cloud Resources, Parallel and Distributed Systems, IEEE Transactions on PP (99) (2014) 1–14. doi:10.1109/TPDS.2014.2355228.

32. Y. Mansouri, A. Toosi, R. Buyya, Brokering Algorithms for Optimizing the Availability and Cost of Cloud Storage Services, in: Cloud Computing Technology and Science (CloudCom), 2013 IEEE 5th International Conference on, Vol. 1, 2013, pp. 581–589. doi:10.1109/CloudCom.2013.83.

33. D. Loreti, A. Ciampolini, A Hybrid Cloud Infrastructure for Big Data Applications, in: Proceedings of the 2015 IEEE 17th International Conference on High Performance Computing and Communications, 2015 IEEE 7th International Symposium on Cyberspace Safety and Security, and 2015 IEEE 12th International Conf on Embedded Software and Systems, HPCC-CSS-ICESS '15, IEEE Computer Society, Washington, DC, USA, 2015, pp. 1713–1718. doi:10.1109/HPCC-CSS-ICESS.2015.140.

34. Z. Zheng, Y. Gui, F. Wu, G. Chen, STAR: Strategy-Proof Double Auctions for Multi-Cloud, Multi-Tenant Bandwidth Reservation, Computers, IEEE Transactions on PP (99) (2014) 1–14. doi:10.1109/TC.2014.2346204.

35. A. Iosup, N. Yigitbasi, D. Epema, On the Performance Variability of Production Cloud Services, in: Cluster, Cloud and Grid Computing (CCGrid), 2011 11th IEEE/ACM International Symposium on, 2011, pp. 104–113. doi:10.1109/CCGrid.2011.22.

36. N. Grozev, R. Buyya, Performance Modelling and Simulation of Three-Tier Applications in Cloud and Multi-Cloud Environments, The Computer Journal 58 (1) (2015) 1–22. doi:10.1093/comjnl/bxt107.

37. B. Sharma, T. Wood, C. Das, HybridMR: A Hierarchical MapReduce Scheduler for Hybrid Data Centers, in: Distributed Computing Systems (ICDCS), 2013 IEEE 33rd International Conference on, 2013, pp. 102–111. doi:10.1109/ICDCS.2013.31.

38. R. Tudoran, K. Keahey, P. Riteau, S. Panitkin, G. Antoniu, Evaluating Streaming Strategies for Event Processing Across Infrastructure Clouds, in: Cluster, Cloud and Grid Computing (CCGrid), 2014 14th IEEE/ACM International Symposium on, Chicago, IL, USA, 2014, pp. 151–159.

39. M. Zaharia, T. Das, H. Li, S. Shenker, I. Stoica, Discretized streams: an efficient and faulttolerant model for stream processing on large clusters, in: Proceedings of the 4th USENIX conference on Hot Topics in Cloud Computing, HotCloud'12, USENIX Association, Berkeley, CA, USA, 2012, pp. 10–10.
40. M. Ding, L. Zheng, Y. Lu, L. Li, S. Guo, M. Guo, More Convenient More Overhead: The Performance Evaluation of Hadoop Streaming, in: Proceedings of the 2011 ACM Symposium on Research in Applied Computation, RACS '11, ACM, New York, NY, USA, 2011, pp. 307–313. doi:10.1145/2103380.2103444.
41. M. Zaharia, A. Konwinski, A. D. Joseph, R. Katz, I. Stoica, Improving MapReduce Performance in Heterogeneous Environments, OSDI (2008) 29–42.
42. B. Tang, H. He, G. Fedak, HybridMR: a new approach for hybrid MapReduce combining desktop grid and cloud infrastructures, Concurrency and Computation: Practice and Experience 27 (16) (2015) 4140–4155.
43. F. J. Clemente-Castello, B. Nicolae, K. Katrinis, M. M. Rafique, R. Mayo, J. C. Fernandez, D. Loreti, Enabling Big Data Analytics in the Hybrid Cloud Using Iterative MapReduce, in: Utility and Cloud Computing - UCC, 2015 IEEE/ACM 8th International Conference on , IEEE Computer Society, 2015, pp. 290–299. doi:10.1109/UCC.2015.47.

Distributed Computing Technologies in Big Data Analytics

Kaushik Dutta

1 Introduction

The database technology has evolved over time. As the application of database has extended from simple mainframe to desktop application to web application to mobile application, the size of data to store and manage through database has also increased. Figure 1 depicts this growth of the data. The first generation of data growth came from ERP software and following that with the introduction of CRM. Next, the introduction of web moved the data volume to terabyte range. However, with the mobile, sensor and social media based applications, the data volume is growing in the range of petabytes.

Relational database (RDBMS) has been one of the most successful database technology since the 1980s. However, even with its solid technological growth, the relational database has failed to scale with the growth of data. Despite the advances in computing, faster processors and high-speed networks, the scalability of the relational database has been restricted. The applications built using RDBMS technology either has failed to perform with increased data or the cost of the infrastructure to keep the application performing has grown exponentially.

Secondly, the relational database was designed for tabular data with a consistent structure and fixed schema. Relational database works best when the structure of the data is known beforehand. However, in the new world as the volume and velocity of the data are increasing, so is the variety and complexity of data. Applications need to be built into the database without the full understanding of the data to be stored and the structure of the data. Or, the structure of the data is being changed after the application has been built. For example, consider a retail application that is selling electronic goods. It can develop applications that can search and manage the known

K. Dutta (✉)
University of South Florida, Tampa, FL, USA
e-mail: duttak@usf.edu

© Springer International Publishing AG 2017
S. Mazumder et al. (eds.), *Distributed Computing in Big Data Analytics*,
Scalable Computing and Communications, DOI 10.1007/978-3-319-59834-5_4

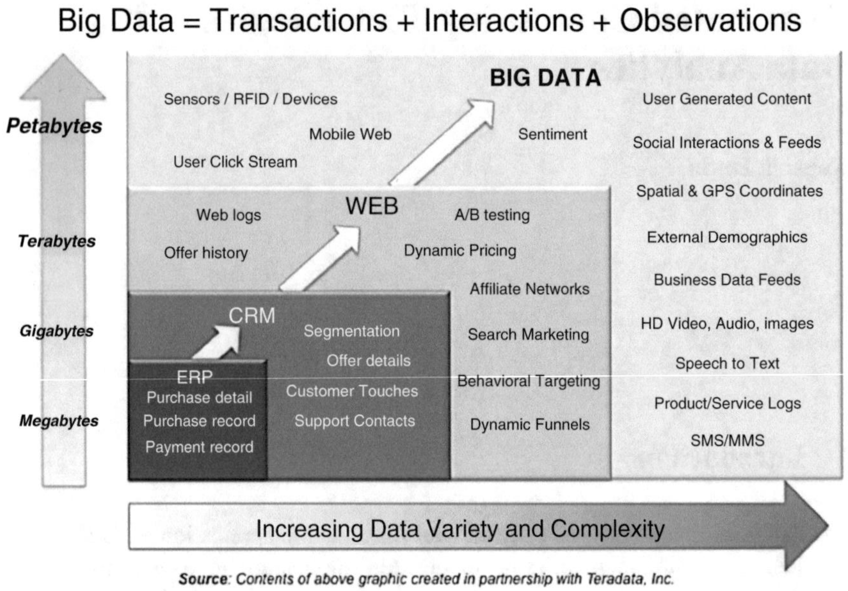

Big Data = Transactions + Interactions + Observations

Fig. 1 Data growth

set of electronic goods. However, if in the future a new device comes up (such a brain-reader) with a new set of features and specifications, the applications will not know how to store and manage that device through a relational database. The relational structure does not allow to handle such unstructuredness with the data.

In the next section, we describe the fundamentals of distributed database and the basics of the no-SQL database. In describing the technologies, we rely on few software platforms supporting such tools. Though there are many different software platforms supporting the similar technologies, mostly we have chosen the software that is popular and preferably supported by open source platforms. The database discussion is followed by the distributed file system and distributed computing platform such as map-reduce and spark. Next, we follow the discussion on how these distributed technologies are being used to develop a newer generation of machine learning platforms. The textual document is one of the important sources of today's information. We describe the basics of textual search platform and associated software such as Lucene and ElasticSearch. Distributed caching enhances the performance of real-time access to data. We describe the distributed caching systems such as REDIS. As a number of components and systems grow exponentially in big data infrastructure, the communication across these components need to be managed more efficiently. In this context, we describe the message passing software such as RabbitMQ and Kafka. Lastly, the traditional tools are unable to represent big data visually. The Newer generation of visualization tools is being developed to present the data. We describe these big data visualization tools in the later part of this chapter.

2   Distributed Database

The scalability issues in the relational database come with the ACID property. The ACID (Atomicity, Concurrency, Isolation, and Durability) property ensures the consistency of data and helps to execute transactions in databases. Database vendors long ago recognized the need for partitioning databases and introduced a technique known as 2PC (two-phase commit) for providing ACID across multiple database instances [35].

It is relatively easy to maintain the ACID property in a single server database system or even with a two node master-slave database server. However, as the data volume grows, it becomes necessary to distribute the data across multiple nodes. With multiple nodes, the cost of communication to maintain ACID property increases. Also, the availability of any system is the product of the availability of the components required for operation. A transaction involving two databases will have the availability of the product of the availability of each database. For example, if we assume each database has 99.9% availability, then the availability of the transaction becomes 99.8%, or an additional downtime of 43 min per month [35].

This leads us to an important barrier of distributed system – Brewer's theorem [12] on the correlation between consistency, availability, and partition-tolerance. Brewer postulates three distinct properties for distributed systems with an inherent correlation [18].

Consistency The consistency property describes a consistent view of data on all nodes of the distributed system. That is, the system assures that operations have an atomic characteristic and changes are disseminated simultaneously to all nodes, yielding the same results.

Availability This property demands the system to eventually answer every request, even in the case of failures. This must be true for both read and write operations.

Partition Tolerance This property describes the fact that the system is resilient to message losses between nodes. And according to the availability property, every node of any potential partition must be able to respond to a request.

The core statement of Brewer's theorem is: *"You can have at most two of these properties for any shared-data system."* (Fig. 2).

Though all the above properties in a distributed database system are desirable, any two of these three properties can be achieved [20].

In a distributed database the data is distributed across multiple geographical sites (as depicted in Fig. 3). In the new era of globalization, distributed database has become a common scenario due to several reasons – (1) Support the distributed Nature of Organizational Units (2) Support the need for Sharing of Data across multiple units and (3) Support for Multiple Application Software. Most of the COTS (commercial off-the-shelf) database server has a distributed version. For example, Oracle has a distributed database since Oracle 7. MySQL Cluster is the distributed version of MySQL Database. IBM's DB2 has several versions of distributed base as part of the DB2 package. However, in the spirit of maintaining

Fig. 2 Different properties
that a distributed system
can guarantee at the same
time (Courtesy: Benjamin
Erb [18])

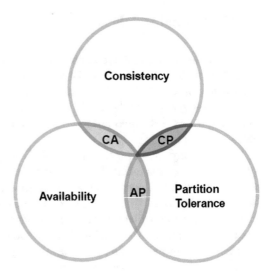

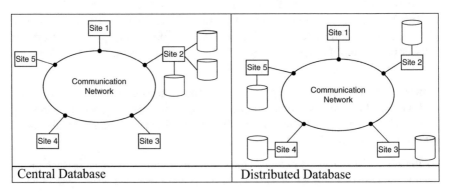

Fig. 3 Distributed database [34]

the ACID property, all these relational models distributed database system has
given up either the partition tolerance or the availability for consistency. When
network separation happens across multiple sites in the distributed relational database, the database fails to serve for that portion of data.

2.1 NoSQL Database

In recent years, a new generation of the database has come up to handle the issues
as discussed above with distributed relational database. It is NoSQL database. As
the name suggests, it does not follow the relational structure – that allows to store
and manage unstructured/semi-structured and unknown data structures. NoSQL
systems are distributed, non-relational databases designed for large-scale data

storage and massively-parallel data processing across a large number of commodity servers. The No-SQL database break through conventional RDBMS performance limits by employing NoSQL-style features such as relaxed ACID property, and schema-free database design. Unlike relational databases, NoSQL database has loosened up the consistency requirements to achieve better availability and partitioning [35].

There are three types of No-SQL databases [29].

Key-Value Stores As the name implies, a key-value store is a system that stores value indexed for retrieval by keys. These systems can hold structured or unstructured data. Typically, these database store items as alphanumeric identifiers (keys) and associated values in simple, standalone tables (referred to as ——hash tables). The values may be simple text strings or more complex lists and sets. Data searches can usually only be performed against keys, not values, and are limited to exact matches [29].

The simplicity of Key-Value Stores makes them ideally suited to lightning-fast, highly scalable retrieval of the values needed for application tasks like managing user profiles or sessions or retrieving product names. This is why Amazon makes extensive use of its Key-Value system, Dynamo, in its shopping cart. Dynamo is a highly available key-value storage system that some of Amazon's core services use to provide highly available and scalable distributed data store [16].

The examples in this category include Amazon's Dynamo [3, 16], Aerospike [2], BerkleyDB (now Oracle No-SQL database) [33] and Riak [11] (Table 1).

Document-Based Stores These databases store and organize data as collections of documents, rather than as structured tables with uniform-sized fields for each record. With these databases, users can add any number of fields of any length to a document. It is designed to manage and store documents. These documents are encoded in a standard data exchange format such as XML, JSON (Javascript Option Notation) or BSON (Binary JSON). Unlike the simple key-value stores described above, the value column in document databases contains semi-structured data – specifically attribute name/value pairs. A single column can house hundreds of such attributes, and the number and type of attributes recorded can vary from row to row. Also, unlike simple key-value stores, both keys and values are fully searchable in document databases [29] (Table 2).

Document databases are good for storing and managing Big Data-size collections of literal documents, like text documents, email messages, and XML documents, as well as conceptual documents like denormalized (aggregate) representations of a

Table 1 Example of key-value store

Product ID (KEY)	Value (Product)
123112	Apple iPhone, 8GB, Gold
146177	Android, Samsung, Galaxy S7, 32GB, US Warranty, Lock Free
123112	Android, Samsung, Galaxy J7, Gold, Dual Sim

Table 2 Example of
document database

{ "ProductID" : "123112",
"Manufacturer": "Apple",
"Model" : "iPhone",
"Memory" : "8GB",
"Color" : Gold}
{ "ProductID" : "146177",
"Manufacturer": "Samsung",
"OS" : "Android",
"Model" : "Galaxy S7",
"Memory" : "32GB",
"Warrantee" : "US Warrantee",
"Lock" : "Lock Free"}
{ "ProductID" : "123112",
"Manufacturer": "Samsung",
"OS" : "Android",
"Model" : "Galaxy J7",
"Color" : "Gold",
"SIM" : "Dual Sim"}

database entity such as a product or customer. They are also good for storing sparse data in general, that is to say, irregular (semi-structured) data that would require an extensive use of nulls in an RDBMS (nulls being placeholders for missing or nonexistent values). The examples of document database are – CouchDB (JSON) [5] and MongoDB (BSON) [32].

Column-Oriented Database These types of database store sets of information in a heavily structured table of columns and rows with uniform-sized fields for each record, as is the case with relational databases, column-oriented databases contain one extendable column of closely related data. It employs a distributed, column-oriented data structure that accommodates multiple attributes per key. While some Wide Column (WC) /Column-Family (CF) stores have a Key-Value DNA (e.g., the Dynamo-inspired Cassandra), most are patterned after Google's Bigtable [13]. Google Bigtable is the petabyte-scale internal distributed data storage system Google developed for its search index and other collections like Google Earth and Google Finance. The tables with column-oriented databases are called column family [29] (Fig. 4).

This type of DMS is great for (1) Distributed data storage, especially versioned data because of WC/CF time-stamping functions. (2) Large-scale, batch-oriented data processing: sorting, parsing, conversion (e.g., conversions between hexadecimal, binary and decimal code values), algorithmic crunching, etc. (3) Exploratory and predictive analytics performed by expert statisticians and programmers. Examples of the Column-oriented database includes Cassandra and SimpleDB.

The Key-Value store databases are completely unstructured. The only query possible in key-value databases is given a key retrieving the value. The document

Product Table (Column Family)				
Row Key: 123112				
ProductID	*Manufacturer*	*Model*	*Memory*	*Color*
123112	Apple	iPhone	8GB	Gold
Row Key: 146177						

ProductID	*Manufacturer*	*Model*	*Memory*	*OS*	*Warrantee*	*Lock*
146177	Samsung	Galaxy S7	32GB	Android	US	Lock Free

Row Key 123112					
ProductID	*Manufacturer*	*Model*	*Color*	*OS*	*SIM*
123112	Samsung	Galaxy J7	Gold	Android	Dual SIM

Fig. 4 Example of column-oriented database

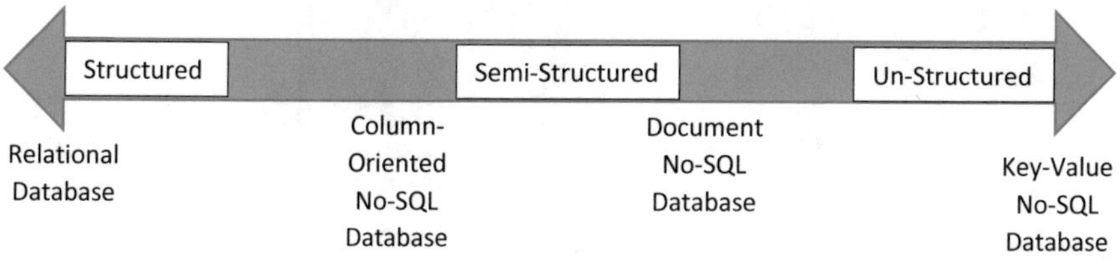

Fig. 5 No-SQL database types

database provides some structure in the value by providing a constraint that the value has to be in JSON or BSON (or any other standard format) format. Other than retrieving the value based on the key, in the document database, it is possible to query based on the content of the value. For example, in MongoDB, the JavaScript based query is used to run a complex query on the value. The column-oriented database has a table structure very similar to a relational database, however, unlike relational tables, the tables with the column-oriented database may have different rows in different columns in the same table. This makes the column-oriented database to handle semi-structured data, where data can be parsed and put into a structured format – but the structure may change from one data item to the next data item. Similar to relational databases, the column-oriented database has high-level query language very similar to SQL. For example, in Cassandra, we have CQL (Cassandra Query Language) [15]. Recently growing number of column-oriented No-SQL databases are implementing SQL-like query capability. Figure 5 depicts the sliding scale of structures in the data and where the different types of No-SQL database fall on this scale. The difference between these different types will be blurred as a growing number of products in one category will incorporate features from other categories.

3   Distributed Storage

Though No-SQL database grew up as a requirement to support the growth of data in volume and variety, not every application requires a database to store and manage the data. Documents, images, videos can be stored in the file system and processed with domain specific tools such as text parser and image processing software. As the size of data captured in these forms (i.e. Documents, images, and videos) are increasing, it became difficult to store and manage it in a single node computing system.

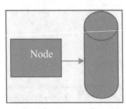

Single node computing with
Single large disk

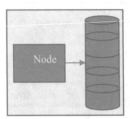

Single node computing with
multiple disks in RAID

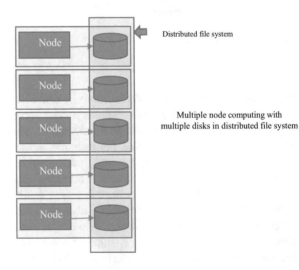

Distributed file system

Multiple node computing with
multiple disks in distributed file system

Traditionally a single node computing system has processed the data stored in local file system. RAID-based storage has come up to accommodate large volumes of data in the file system. The RAID also provides failover mechanism. However, still, only single node can process the data stored in RAID. Processing large volume of data in single node system has been almost impossible, just reading Terabytes of data from a hard disk by a single node computing machine will take several days – any processing on that data will increase that time considerably. To solve this issue the distributed file system has been developed, where the data is distributed across multiple local hard disks each associated with a separate computing node. In such a system, if the computation on the data in the distributed file can be divided in such a way that each node does the processing on the data stored on its local hard disk and the processing in each of these nodes can be done in parallel – then we can complete the processing of terabytes of data in few minutes with the help several hundreds of such nodes. This motivated the development of distributed file system (such as HDFS (Hadoop distributed file system), GFS (Google File System), Amazon S3) and corresponding programming framework map-reduce and Spark. In this section, first we will discuss the distributed file system HDFS, then we will discuss the map-reduce programming framework, and lastly, we will discuss the Spark.

3.1 Hadoop Distributed File System (HDFS)

The HDFS is a distributed file system that spans across multiple nodes. Each of these nodes will have a local regular operating system (such as Linux), on top of which the HDFS file system is deployed. The interface to HDFS is patterned after the UNIX file system (Fig. 6).

HDFS store file system metadata and application data (i.e. the actual files) separately. It stores metadata on a dedicated server called NameNode. Application data are stored on other servers called DataNodes [39]. The DataNode in HDFS does not have any individual failover mechanism such as RAID. Rather the file content is replicated on multiple DataNode for reliability. This has the advantage of data being local to the node, where the computation will be carried out. This reduces the overhead associated with data transfers between the nodes for computational requirements. The GFS [19] has the similar structure (Fig. 7).

The HDFS namespace is a hierarchy of files and directories. File and directories are represented on the NameNode by inodes, which record attributes like permissions, modifications and access times, namespace and disk space quotas. The file content is split into blocks. Conceptually, this is very similar to regular file system blocks, but are much larger in size – typically 128 MB, but may be larger as selected by the user. These are HDFS blocks. Each HDFS block is replicated at multiple DataNodes. The NameNode maintains the namespace tree and the mapping of the HDFS blocks to DataNodes (the physical location of the HDFS block) [39] (Table 3, Fig. 8).

An HDFS client first contacts the NameNode for the locations of data blocks comprising the file and then reads block contents from the DataNode closest to the

HDFS				
Linux (OS)	Linux (OS)	Linux (OS)	Linux (OS)	Linux (OS)
Node	Node	Node	Node	Node

Fig. 6 HDFS on top of Linux

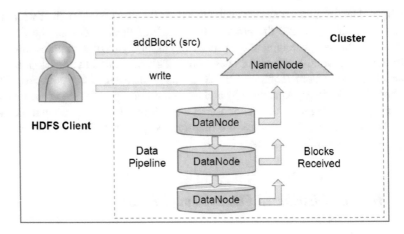

Fig. 7 HDFS architecture [39]

Table 3 NameNode metadata example

Filename	Number of replicas	Block-IDs
/usr/hue/test.dat	3	1, 3, 4, 6
/usr/hue/test2.dat	4	2, 5, 8, 9, 10

Block-ID	Location (DataNode) (Total number of DataNode = 10)
1	1, 3, 5
3	2, 4, 6
4	3, 5, 7
6	4, 6, 8
2	1, 3, 5, 7
5	2, 4, 6, 8,
8	3, 5, 7, 9,
9	4, 6, 8, 10

client. When writing the data, the client requests the NameNode to nominate a suite of DataNodes to host the block replicas. The client then writes data to the DataNodes in a pipeline fashion. HDFS keeps the entire namespace in RAM.

Unlike conventional file system, HDFS provides an API that exposes the locations of a file block. This allows distributed programming like Map-Reduce framework to process data in a node locally where the data is located [39].

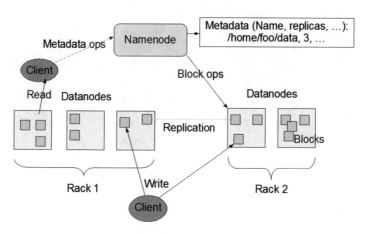

Fig. 8 HDFS architecture (Courtesy: Hortonworks Inc. [23])

4 Distributed Computation

Traditional parallel and distributed computation relied on synchronization and locking. However, the overhead of synchronization across multiple processes and locking the data has considerable overhead. Additionally, the traditional parallel and distributed computation have looked at the computation separately from the data. The assumptions that were made is the data resides in a database or any storage system that is equally accessible by multiple computing nodes. The parallel processing in these nodes will lock the data and process it. In addition to the overhead of locking such an approach adds tremendous overhead in transporting data from the data node (where the data is) to the processing node (the node that is processing the data). The Map-Reduce framework is a new parallel programming framework that addresses these issues with parallel and distributed computing. The Map-Reduce framework is based on two principles.

1. If the computation can be divided based on data segmentation, such that each computational node is processing a different part of the data, the requirement of lock and synchronization can be avoided. This will improve the performance of the parallel computation.
2. If the computation node processes data that is local to its node, then the overhead of data transmission from data node to computation node can be avoided.

Though the map-reduce framework was first developed as part of the Hadoop ecosystem along with the HDFS, the framework in generic and is applicable in a wider variety of data storage including No-SQL databases such as Cassandra and MongoDB.

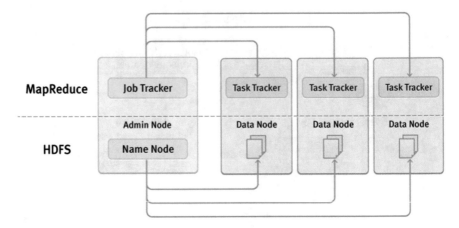

Fig. 9 Map-Reduce on HDFS (Courtesy: NDM Technologies)

4.1   Map-Reduce in Hadoop

Figure 9 depicts the map-reduce architecture on HDFS. The Job Tracker in Map-Reduce is responsible for breaking the job into multiple tasks and assigning to various nodes. The Task Trackers are responsible for completing a task. The Job Tracker and The Name Node of HDFS can coexist in the same node. The Task Tracker and the HDFS Data Node coexist in the HDFS Hadoop framework. Such coexistence allows the Task Tracker to process local data without transmitting the data from one node to another node. The Job Tracker distributes the jobs in such a way that task tracker processes only the local data as far as possible. In Hadoop 2.0 replaced the Job Tracker with Yarn, a separate software component to manage the tasks.

The programming framework of Map-Reduce is based on considering data not as a single unit, but as a collection of multiple units. The example of such collection is – a file is a collection of lines, a directory is a collection of files, a database table is a collection of multiple rows and so on. In Map-Reduce term this collection is considered as a map. Thus the input to map-reduce programming is a map. For example, if a map has N units and there are m task trackers, the Job Tracker can ideally provide N/m units to each task tracker to complete. Obviously, the division of tasks across task-tracker will seldom be so uniform due to non-uniform distribution of data across DataNodes (Fig. 10).

A map-reduce programming framework works in three steps.

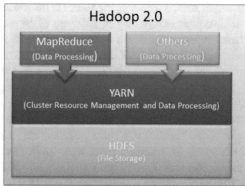

Fig. 10 Hadoop 1.0 vs. Hadoop 2.0 (Courtesy: Saphanatutorial [38])

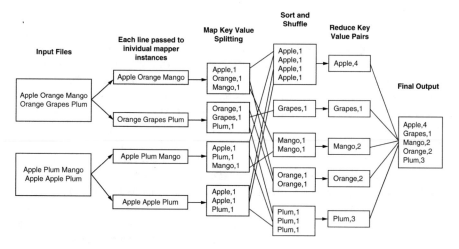

Fig. 11 Map-Reduce example (Source: kickstarthadoop [28])

The input to Map step is set of (Key$_{map-input}$, Value). The output of map step is another set of (key$_{map-output}$, value). The shuffle and sort step sorts the output of map based on keys, group them together and send it to the reducer. So, the reducer input is another map of the form (key$_{map-output}$, {Value, Value, ... , Value}), where each Key of the reducer is associated with a set of values coming out of the map step against that key. Note that the output keys of Map step are the same as the input keys of the reduce step.

Figure 11 provides an example of a map-reduce way of doing distributed computing to compute the word-count distribution in a group of files. The input to the map step is a set of files. Each file is split as a collection of lines. The collection of all the lines is a map, where the key is the location of the line and the value is the line. This map is the input of the Map step in this program. In most cases the map-reduce program does not use input keys, it uses only the values in the input map. The map step involves initiating multiple programs (in different threads in the same node and multiple nodes), each of these programs is called mapper. The input to each mapper is one entry of the input collection, i.e. the map of lines. Thus the input of each mapper in Fig. 11 is a line. The mapper program splits the line into words and creates a map of (Word, 1), where 1 is the count of the word in that line. The mapper program sorts and shuffles this map. With the help of sorts and shuffle, all the mapper programs send the entries (i.e. the count) associated with the same key (i.e. same word) to the same reducer. The reducer upon receiving all the values associated with a key (word, {count, count,..., count}) sums up all the counts for that word and writes into HDFS. Each reducer writes its output independently as a separate file in the HDFS. This creates multiple output file of a map-reduce program running on top of HDFS.

4.2 Spark

In reality, for big data, a single map-reduce program cannot complete the computation required out of the data for analytic purposes. In most cases, a realistic data analytic computation requires a series of map-reduce programs. For example, computing the mean in a map-reduce form can be done in a single map-reduce program. However, the computation of the standard deviation in a map-reduce form will require two sequential map-reduce programs. The first map-reduce computation will compute the mean. Using the result of the first map-reduce, the second map-reduce computation will compute the standard deviation. The input to each map-reduce program is taken from HDFS or some other distributed persistent storage (such as No-SQL database). The output of each map-reduce program is also written into HDFS. This is depicted in Fig. 12.

However, the above workflow will be slow and time-consuming due to multiple reads and write from the HDFS system. Additionally, though map-reduce was developed to run large distributed parallel computing process on a number of regular consumer hardware, in present days computing machines with higher memory and processing capacity is very common in enterprise architecture. The spark has been developed to use the larger memory capacity of today's computing hardware.

Spark exploits the memory capacity to avoid the repeated reading and writing on the map-reduce workflow. Spark has a concept called "Resilient Distributed Data" (RDD). In the most simplistic concept, the RDD considers the memory across

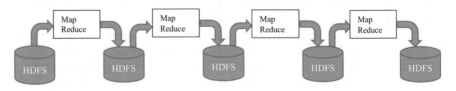

Fig. 12 Map-Reduce workflow

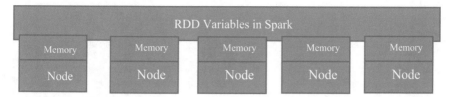

Fig. 13 RDD in spark

multiple computers as a single contiguous memory. Typical RDD variables are collections (such as map, array, list) that stay in memory but spans across computer boundary. This results in two advantages. First, as the collection is distributed across multiple machines, any processing of the collection can be done in parallel on all these machines, where each machine does the computation on its local memory (very similar to map-reduce computation). Second, the distributed map-reduce processing in the case of the spark is done on RDD (memory resident collections), so the processing is much faster than the HDFS based map-reduce (Fig. 13).

In the case of RDD in Spark, the map is a transformation that passes each item in the RDD through a function and returns a new RDD representing the result. For example, there is an RDD x a collection of 10 K integers. We want to increase each item in the RDD by 1. This can be carried out as an RDD map. The reduction on an RDD is an action that aggregates all the elements of the RDD use some function and returns the final result. The example of reducing on x will be sum all the values. Typically RDDs are kept in the memory and cease to exist once the spark program execution has finished. However, it is also possible to persist an RDD in memory, in which case the Spark will keep the elements around on the cluster for much faster access the next time we carry. There is also support for persisting RDDs on disk or replicated across multiple nodes.

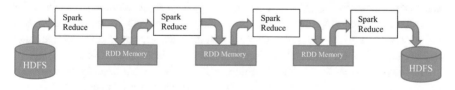

5 Machine Learning Platforms

With the popularity of Spark, running machine learning algorithms on big data has become much easier. The landscape of machine learning on big data in changing dramatically with Spark. All latest machine learning platforms are using Spark in some way or other. Using these platforms companies can build models on large data sets without sampling and achieve accurate predictions. These tools use few optimizations to achieve so. First, they use more memory and processing power for

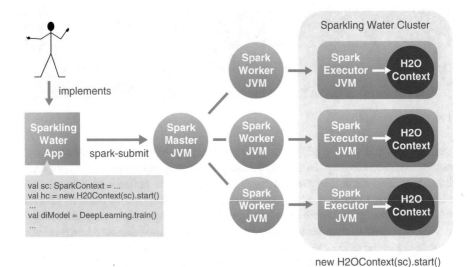

val sc: SparkContext = ...
val hc = new H2OContext(sc).start()
...
val diModel = DeepLearning.train()
...

new H2OContext(sc).start()

Fig. 14 Sparking water architecture (Courtesy: Cloudera [14])

making faster computations. Second, they use in-memory compression to handle large datasets. And third, they implement parallel distributed network training. The deep learning approaches used by these tools build hierarchies of hidden features that is composed to approximate complex functions with much less effort.

The Mahout [6] has been a very popular machine learning platform on HDFS platform. However, as Spark became popular many of the Mahout machine learning libraries migrated to Spark environment.

H2O [22] is another machine learning platform that can work on both Hadoop and Spark. SparkFlows [40] is a Big data application development platform for building and executing end-to-end data analytic products on Spark. It comes pre-packaged with an exhausting set of machine learning and ETL components making the workflow definition of big data use cases faster and easier.

The Sparkling Water project combines H2O machine-learning algorithms with the execution power of Apache Spark. Figure 14 illustrates the concept of technical realization. The application developer implements a Spark application using the Spark API and Sparkling Water library.

6   Search System

With the growth of data, the requirement of real-time delivery of information has grown also. This has particularly become true for textual data. A vast amount of big data is unstructured textual data, such as the posts derived from Twitter, Facebook, and blogs, or textual description of products, or archival data of legal documents.

ID	Text
1	Baseball is played during summer months.
2	Summer is the time for picnics here.
3	Months later we found out why.
4	Why is summer so hot here
↑	Sample document data

Dictionary and posting lists →

Term	Freq	Document ids
baseball	1	[1]
during	1	[1]
found	1	[3]
here	2	[2], [4]
hot	1	[4]
is	3	[1], [2], [4]
months	2	[1], [3]
summer	3	[1], [2], [4]
the	1	[2]
why	2	[3], [4]

Fig. 15 Inverted index example (Source: Hotcodeshare [25])

The storage consisting of such textual data can easily reach in the range of few hundreds of terabytes to a petabyte. The real-time search of this data is impossible to achieve with the traditional database indexing scheme.

To make the textual data searchable, an inverted index is created out of textual data. In forward index, a document is stored in the database, and with a document ID, we can retrieve the document. An inverted index an index is created by words in documents. Then each word in the index points to the set of the documents that contain that word. Figure 15 shows one example of the inverted index.

The first table in Fig. 15 is the data of documents along with the forward index; the second right-hand side table is the inverted index. With the second table, one can easily answer queries such as "Find all the documents containing the word 'summer.'" Without the inverted index, such query would have taken a long time by searching for the word `summer' in each and every document on the table. As the number of words in a language is limited, even with a very large number of documents the number of entries in the inverted index will be limited, and thus the makes it possible to hold the index in memory of a single node or a cluster of nodes. The basic algorithm of the inverted index was implemented as part of Lucene library [9].

6.1 Search Software

Solr [4] was developed on top of Lucene to have a server version that can support HTTP and XML based query. With big data, the requirement evolved to hold the index larger than single machine memory and have replication of the index to accommodate failover of a node hosting the index. This resulted in the development of ElasticSearch [17].

Elasticsearch is a distributed, RESTful search engine. It supports HTTP and JSON based query capability. Though the basics of ElasticSearch evolved to host the inverted index of textual data, the ElasticSearch can host index of any data. Say,

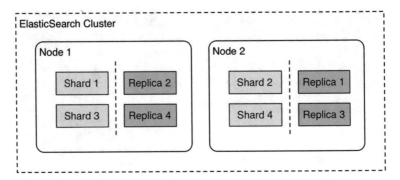

Fig. 16 Failover and clustering in ElasticSearch (Source: Liip [30])

for example; it can hold the index of product attributes such as manufacturer, model, price, year, rating, keywords in the product title and keywords in product descriptions. Typically ElasticSearch is not used to store the actual data; it is used to store memory-resident index structure that can search and queried in real-time. In big data architecture, it is very common practice to query the ElasticSearch to retrieve the document ID (such as product ID) and then to query the HDFS or No-SQL database to retrieve the actual document (or product details).

As shown in Fig. 16, the ElasticSearch has the inbuilt replica and sharding structure. The sharding allows the single index to be broken down into multiple partitions in different nodes. Each shard can be queried in parallel to retrieve data against a single query. This improves the query performance in ElasticSearch. Secondly, the sharding also allows the index larger than a single node memory to be stored and managed by ElasticSearch. The replica in ElasticSearch improves the reliability and failover mechanism in ElasticSearch making it a search platform of choice for online real-time applications.

Message Passing and Queuing System

In a big data system, nothing is a single node system – every component is a cluster of a large number of nodes that handle the distributed data and computation. In such a scenario, creating and managing one to one communication becomes a challenge. Consider a scenario where the data is coming from multiple sensors. A process is receiving the data from sensors and processing it to identify the structured data, and writing the semi-structure data into various data storage depending on the type of data and information (Fig. 17).

One of the critical problems with the above architecture is the flow of data out of sensor is non-uniform making it difficult to estimate the infrastructure requirement for processing the data. There will be a mismatch in the rate at which the data is coming out of sensors and the rate at which the data can be processed to write into the storage. This will result in having a large in-memory buffer in the data parsing and extraction program. Additionally, in case the data parsing and extraction program fail during processing, there will be a loss of data.

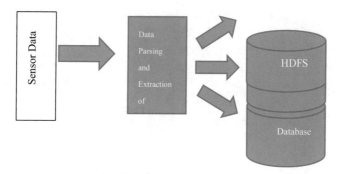

Fig. 17 Data processing & storing workflow

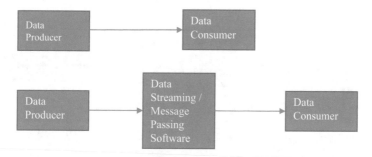

Fig. 18 Role of data streaming/message passing software

7 Big Data Messaging Software

To handle the above issues, a class of software has evolved – called message passing or stream processing software such as RabbitMQ, Kafka, Kinesis, Flink. These software components allow handling a large volume of messages. These software has the capability to hold the messages temporarily with failover and replication capability and can process the messages before passing it to the data consumer (Fig. 18).

RabbitMQ [36] is one of the leading message passing software that has been popular in IT infrastructure to manage streaming data since pre-big-data days. Traditionally RabbitMQ is a single server system, thought with the growing popularity of big data it has incorporated clustering in its architecture. The RabbitMQ has the capability to incorporate complex routing logic based on message content. The most popular message passing system in RabbitMQ is the pub-sub system. In the pub-sub system, a group of message producers publishes messages with subjects and a group of consumers consume these messages based on the subjects (Fig. 19).

Apache Kafka [8] is a clustered stream data processing software. Unlike RabbitMQ which can typically process messages in a range of 20–30 K per seconds, with the inbuilt clustering technology a Kafka cluster can process a much

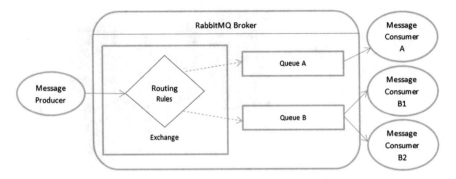

Fig. 19 Messaging with RabbitMQ (Source: https://keyholesoftware.com/2013/05/13/messaging-with-rabbitmq/)

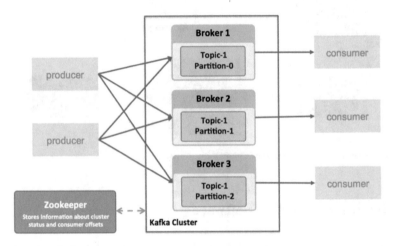

Fig. 20 Kafka cluster (Source: Hortonworks [24])

higher number of messages (100 K to few million messages per seconds). A Kafka cluster consists of multiple partitions and multiple servers. Each partition has one server which acts as the "leader" and zero or more servers which act as a "followers." The leaders handle all read and write requests for that partition while the followers passively replicate the leader. If the leader fails, one of the followers will automatically become the new leader. Each server acts as a leader for some of its partitions and a follower for others, so the load is balanced across multiple servers. Unlike RabbitMQ whose strength is in routing, the strength of Kafka can consume the massive volume of stream data (Fig. 20).

Apache Flink [7] is a streaming data processing system. It can handle large-scale system running thousands of nodes. It provides accurate computational results on streaming data. A very common use case for Apache Flink is analytics on stream data. Quite often Flink and Kafka are used together, where data streams for Flink are ingested from Kafka. Typically applications of Flink and Kafka start with event

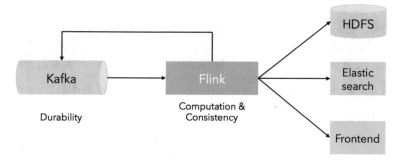

Fig. 21 Kafka and Flink together (Source: [42])

streams being pushed to Kafka, which are then consumed by Flink jobs. These jobs range from simple transformations of data import/export to more complex applications that aggregate data in windows. The results of these Flink jobs may be fed back to Kafka for consumption by other services or written out to other systems like HDFS, Elasticsearch, No-SQL database or web front end. In such a system, Kafka provides data durability, and Flink provides consistent data movement and computation (Fig. 21).

8 Cache

Caching is an important component of any big data-based systems that expect to provide a real-time response to requests. The generic idea of caching is most frequently accesses data items are brought near to the application so that frequent requests of these data items can be served in near real time. In the most application, there is exists a skew in access pattern to data. For example, following power law [1], 80% of the users will access 20% of the data item. These 20% data can be brought into a memory based caching system, from where the requests for the data can be server much faster than persistent storage such as database or file system.

8.1 Distributed Caching Systems

Memcached [31] is a very popular high-performance and distributed memory caching system. In essence, it is an in-memory key-value store for small chunks of data (strings, objects) from results of database calls, file-read or remote service call. Memcached is inbuilt in latest versions of MySQL to cache database calls. Traditionally Memcached has been a single server software component, but with the big data systems, Memcached has also grown to accommodate multi-node

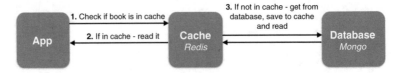

Fig. 22 REDIS cache in front of MongoDB No-SQL database (Source: Gino [21])

Fig. 23 Example architecture for big data analytics

cluster. Redis [37] is another software product that provides cache service. It is an in-memory data structure store and can store many complex objects such as arrays, sets, and lists. Unlike Memcached (which provides get and set operations only) Redis allows atomic operations on these objects such as appending to a string, pushing an element to a list, computing set union, intersection and difference or finding the item in a sorted set. Redis has built-in replication, high availability and data partition feature. Though Redis works with an in-memory dataset, it can persist the data by periodically dumping the data to disk. Redis can be used as both in-memory no-SQL Database and cache (Fig. 22).

Case Study: Big Data Analytics Example Architecture

Figure 23 presents an example architecture of a big data analytics in an organization. At the left-hand side, a massive stream of sensor data and social media data is coming as input. This data is passed to Kafka for temporary holding. The Flink consumes these messages from Kafka, processes and parses it, and writes to appropriate storage (HDFS or No-SQL database). The analytic programs run on top of the data that is stored in HDFS or No-SQL database. These analytics programs can be in the form of mapreduce, spark and use advanced machine learning libraries such as SparkML, SparkFlow, and TensorFlow. The output of analytics program is again saved into the database. Depending on the use case and the data volume, the analytic output can be stored in HDFS, No-SQL database or even in a relational database. The Elasticsearch or equivalent search system is populated with the output of Flink and analytics program for indexing purpose. End users use a front end application to search and retrieve information. The front application is running in the application server first contacts the Elasticsearch to search for the information (such as a list of product ID) based on various attribute values. Then it contacts the database to retrieve the actual information (such as product details) associated with the search results. The cache (such as Redis) may be deployed in between the front end application (in application server) and the data storage to improve the speed of access to popular items (such as daily hot products).

9 Data Visualization

With the increase of data, the visualization of this volume and variety of data has become a challenge. Some tools have emerged in recent years to present the data in innovative ways. Tableau [41] has become a popular technology to do data visualization. There are some tools on visualization that works in the cloud and others that work as a desktop application with cloud-based access to reports. The traditional technologies like Tableau replied on the later. Tableau is primarily used to develop dashboard. Tableau is an end user-friendly tool. As soon as the data is connected with the Tableau, the Tableau GUI can be used to develop various GUI based reports and dashboards.

Recently a new technology "Notebook" has come up as a way to develop and maintain rich visualization of data. One of such software is Jupyter [27]. The notebook in Jupyter contains both computer code and rich text elements (paragraphs, equations, figures, links, etc.). These documents contain analysis description and the results along with the executable code which can be run to perform data analysis. This allows automatic generation of rich text document containing data analysis text and visual representation of the data. The executable code associated with a notebook can be shared and can be modified to develop new reports. In the past, the report generators and reports were two different component in the enterprises that have been maintained separately. This used to lead to a lot of mismatch in report generation code and the actual reports. The Notebook technology allows these two to be merged and considered as a single unit. IBM's Data Science Experience [26] is another such technology by IBM. Apache Zeppelin [10] is an Apache software that supports the Notebook functionality.

10 Conclusion

In this chapter we have discussed various technologies related to big data technology – NoSQL database, distributed file system, map-reduce and spark based distributed computation, distributed communication platform, distributed caching, search platform and visualization technologies. Our intention here was to give an overview of all these technologies so that appropriate technical discussion can be led in future by the readers. Later in the chapter, we delved down into some associated technologies such as search system, message processing and caching that makes the big data analytics application more robust and performance efficient. Lastly, we present an example architecture of a big data analytics based application

Table 4 Summary of big data technology

Technology type	Purpose	Product example	Use case
Database	Distribute data with replication and failover	No-SQL databases such as Cassandra, MongoDB, BerkleyDB, CouchDB, SimpleDB, DynamoDB	Store and manage large volume of structured and unstructured data
File system	File system distributed across multiple nodes	HDFS, GFS	Store and manage large files or a large number of small files
Programming	Distributed programming that can process data in parallel	Hadoop Map-Reduce, Spark	Computation and analytics job on top of data in distributed file system or no-SQL database
Machine learning platform	Complex analytical work using machine learning techniques	Mahout, H2O, SparkML, Sparking Water, SparkFlows	Deep learning and machine learning on big data
Search system	To search unstructured and semistructured data	ElasticSearch, Solr	Store and manage index on big data for search purposes.
Messaging system	To introduce an intermediate buffer between data collection and data storage	RabbitMQ, Kafka, Kinesis, Flink	Read high throughput incoming data (such as twitter data, sensor data) and preprocess it before writing into data storage
Caching	Distributed application level caching	REDIS, Memcached	Store frequently accessed data from in-memory distributed cache to reduce the access time to this data compared to accessing from persistent storage of database or filesystem.
Data visualization	Provides Notebook functionality where report generation code and the actual report co-exists	Tableau, Jupyter, Zeppelin, Data Science Experience	Generate complex report including visual representation and textual description of data

using these technologies. In Table 4 we summarize the technologies and products described in this chapter.

Though there is abundance of technology and software platforms to process, manage and use big data, the appropriate choice is very critical for the success of these platforms. We expect more technology to evolve in next few years to support distributed computation on big data. Once such technology is cryptocurrency (such as bitcoin) and blockchain. The blockchain technology is gradually getting traction to store the data in a peer to peer fashion without the control of any single entity. The technology is now being applied in wide variety of domains including financial, healthcare and contract management. IOT (internet of things) is another distributed technology that is coming up. The application of IOT in every part of our life is becoming the norms, where the data gathering, communicating and processing are interconnected and distributed to the point where data is being generated.

References

1. Adamic, L. A., & Huberman, B. A. (2000). Power-Law Distribution of the World Wide Web. *Science*, *287*(5461).
2. Aerospike. (2017). Aerospike | High Performance NoSQL Database. Retrieved March 23, 2017, from http://www.aerospike.com/
3. Amazon. (2017). AWS | Amazon SimpleDB – Simple Database Service. Retrieved March 23, 2017, from https://aws.amazon.com/simpledb/
4. Apache. (2015). Solr. Retrieved from http://lucene.apache.org/solr/
5. Apache. (2017a). Apache CouchDB. Retrieved March 23, 2017, from http://couchdb.apache.org/
6. Apache. (2017b). Apache Mahout: Scalable machine learning and data mining. Retrieved March 23, 2017, from http://mahout.apache.org/
7. Apache Flink. (2017). Apache Flink: Introduction to Apache Flink®. Retrieved March 23, 2017, from https://flink.apache.org/introduction.html
8. Apache Kafka. (2017). Apache Kafka. Retrieved March 23, 2017, from https://kafka.apache.org/intro
9. Apache Lucene. (2017). Apache Lucene – Welcome to Apache Lucene. Retrieved March 23, 2017, from https://lucene.apache.org/
10. Apache Zeppelin. (2017). Zeppelin. Retrieved March 23, 2017, from https://zeppelin.apache.org/
11. Basho. (2017). Riak – Distributed Databases. Retrieved March 23, 2017, from http://basho.com/products/
12. Brewer, E., & Eric. (2010). A certain freedom. In *Proceeding of the 29th ACM SIGACT-SIGOPS symposium on Principles of distributed computing – PODC '10* (pp. 335–335). New York, New York, USA: ACM Press. http://doi.org/10.1145/1835698.1835701
13. Chang, F., Dean, J., Ghemawat, S., Hsieh, W. C., Wallach, D. A., Burrows, M., … Gruber, R. E. (2006). Bigtable: A distributed storage system for structured data. In *7th Symposium on Operating Systems Design and Implementation (OSDI '06), November 6–8, Seattle, WA, USA* (pp. 205–218). USENIX Association. Retrieved from http://research.google.com/archive/bigtable-osdi06.pdf
14. Cloudera. (2017). How-to: Build a Machine-Learning App Using Sparkling Water and Apache Spark – Cloudera Engineering Blog. Retrieved March 23, 2017, from http://blog.cloudera.com/blog/2015/10/how-to-build-a-machine-learning-app-using-sparkling-water-and-apache-spark/
15. Datastax. (2017). Introduction to Cassandra Query Language. Retrieved March 23, 2017, from https://docs.datastax.com/en/cql/3.1/cql/cql_intro_c.html
16. DeCandia, G., Hastorun, D., Jampani, M., Kakulapati, G., Lakshman, A., Pilchin, A., … Vogels, W. (2007). Dynamo. In *Proceedings of twenty-first ACM SIGOPS symposium on Operating systems principles – SOSP '07* (p. 205). New York, New York, USA: ACM Press. http://doi.org/10.1145/1294261.1294281
17. Elastic. (2017). Open Source Search Analytics · Elasticsearch. Retrieved March 23, 2017, from https://www.elastic.co/
18. Erb, B. (2016). The Challenge of Distributed Database Systems. Retrieved March 23, 2017, from http://berb.github.io/diploma-thesis/community/061_challenge.html
19. Ghemawat, S., Gobioff, H., & Leung, S.-T. (2003). The Google file system. *ACM SIGOPS Operating Systems Review*.

20. Gilbert, S., & Lynch, N. (2002). Brewer's conjecture and the feasibility of consistent, available, partition-tolerant web services. *ACM SIGACT News, 33*(2), 51. http://doi.org/10.1145/564585.564601
21. Gino, I. (2017). Caching a MongoDB Database with Redis — SitePoint. Retrieved March 23, 2017, from https://www.sitepoint.com/caching-a-mongodb-database-with-redis/
22. H2O. (2017). H2O.ai. Retrieved March 23, 2017, from https://www.h2o.ai/h2o/
23. Hortonworks. (2017a). Apache Hadoop HDFS – Hortonworks. Retrieved March 23, 2017, from https://hortonworks.com/apache/hdfs/#section_2
24. Hortonworks. (2017b). Introduction to Kafka – Hortonworks Data Platform. Retrieved March 23, 2017, from https://docs.hortonworks.com/HDPDocuments/HDP2/HDP-2.3.2/bk_kafka-user-guide/content/ch_using_kafka.html
25. Hotcodeshare. (2017). How Elasticsearch index document? | Hot code share. Retrieved March 23, 2017, from http://www.hotcodeshare.com/content/how-elasticsearch-index-document
26. IBM. (2017). IBM Data Science Experience. Retrieved March 23, 2017, from https://www.ibm.com/us-en/marketplace/data-science-experience/resources
27. Jupyter. (2017). Project Jupyter. Retrieved March 23, 2017, from http://jupyter.org/
28. kickstarthadoop. (2017). Kick Start Hadoop: Word Count – Hadoop Map Reduce Example. Retrieved March 23, 2017, from http://kickstarthadoop.blogspot.com/2011/04/word-count-hadoop-map-reduce-example.html
29. Leavitt, N. (2010). Will NoSQL Databases Live Up to Their Promise? *Computer, 43*(2), 12–14. http://doi.org/10.1109/MC.2010.58
30. Liip. (2017). On ElasticSearch performance – Liip Blog. Retrieved March 23, 2017, from https://blog.liip.ch/archive/2013/07/19/on-elasticsearch-performance.html
31. Memcached. (2017). memcached – a distributed memory object caching system. Retrieved March 23, 2017, from https://memcached.org/
32. MongoDB Inc. (2015). mongoDB. Retrieved from https://www.mongodb.org/
33. Oracle. (2017). Berkeley DB Products. Retrieved March 23, 2017, from https://www.oracle.com/database/berkeley-db/index.html
34. Ozsu, M. T., & Valduriez, P. (2011). *Principles of Distributed Database Systems – M. Tamer Özsu, Patrick Valduriez – Google Books*. Retrieved from https://books.google.com/books?hl=en&lr=&id=TOBaLQMuNV4C&oi=fnd&pg=PR3&dq=Distributed+Database&ots=LqFjgM_P-7&sig=mcmEnxerBLtixHY-0CrzS2hFojc#v=onepage&q=Distributed Database&f=false
35. Pritchett, D., & Dan. (2008). BASE: AN ACID ALTERNATIVE. *Queue, 6*(3), 48–55. http://doi.org/10.1145/1394127.1394128
36. RabbitMQ. (2017). RabbitMQ – Messaging that just works. Retrieved March 23, 2017, from https://www.rabbitmq.com/
37. Redis. (2017). Redis. Retrieved March 23, 2017, from https://redis.io/
38. Saphanatutorial. (2017). How YARN Overcomes MapReduce Limitations in Hadoop 2.0. Retrieved March 23, 2017, from http://saphanatutorial.com/how-yarn-overcomes-mapreduce-limitations-in-hadoop-2-0/
39. Shvachko, K., Kuang, H., Radia, S., & Chansler, R. (2010). The Hadoop Distributed File System. In *2010 IEEE 26th Symposium on Mass Storage Systems and Technologies (MSST)* (pp. 1–10). IEEE. http://doi.org/10.1109/MSST.2010.5496972
40. Sparkflows. (2017). SparkFlows.io | Big Data Application Development Made Easy. Retrieved March 23, 2017, from https://www.sparkflows.io/overview
41. Tableau. (2017). Business Intelligence and Analytics – Tableau Software. Retrieved March 23, 2017, from https://www.tableau.com/
42. Tzoumas, K., & Metzger, R. (2015). Kafka + Flink: A practical, how-to guide – data Artisans. Retrieved March 23, 2017, from https://data-artisans.com/kafka-flink-a-practical-how-to/

Security Issues and Challenges in Big Data Analytics in Distributed Environment

Mayank Swarnkar and Robin Singh Bhadoria

1 Introduction

With the growth of technology, there is a huge expansion in the data generation and its exchange over the Internet. This growth is massive and hence tough for traditional processing systems to process this amount data. Every day more than 2.5 Quintilian bytes of data is generated [1, 2]. Ninety percent of total data in the world is generated in past 2–3 years [3]. This huge amount of data is termed as big data. Traditionally, when it is not feasible to process data in a single machine, we take help of distributed systems [23]. Multiple systems together process data in parallel or sequentially using distributed processing algorithms [24]. But to process big data, traditional algorithms of distributed processing systems are not efficient [4]. Therefore, to process big data, many big data processing tools are designed. Few of the famous big data processing tools are Hadoop [5, 6], Spark [7], Hive etc.

Evolution of big data brings many security issues with it. Traditional security mechanisms are designed for securing small and static data-sets. Those security mechanisms are unfit for big data. Big data is efficiently processed in distributed environments instead of on single machine. Algorithms like Map-Reduce [25] and SCOPE [26] became the base of big data processing in distributed environment. The key to processing big data is to divide the data into chunks and give it to multiple processing units or nodes in distributed environment. When the data gets processed,

M. Swarnkar (✉)
Indian Institute of Technology, Indore, Madhya Pradesh, India
e-mail: swarnkar.mayank@gmail.com

R.S. Bhadoria
Discipline of Computer Science and Engineering, Indian Institute of Technology Indore,
Indore - Khandwa Road, Indore, Madhya Pradesh, India
e-mail: robin19@ieee.org

© Springer International Publishing AG 2017
S. Mazumder et al. (eds.), *Distributed Computing in Big Data Analytics*,
Scalable Computing and Communications, DOI 10.1007/978-3-319-59834-5_5

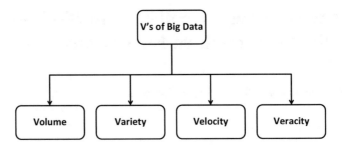

Fig. 1 V's of big data

combine the result and give output. This imports another security issues in big data. Now, security issues of big data and distributed systems needs to be handled together.

In this chapter, we are going to discuss security issues in big data in distributed environment. we give an overview of various security aspects of big data in distributed environment. We also discuss some available solutions researched in literature.

Figure 1 shows the V's of big data. These V's are the drivers of Big Data. Data is converted to big data when it follows the properties mentioned as V's of big data. We discuss another view of these drivers of big data. We say that these V's are not only drivers of big data but also the up-bringer of security issues in data analytics. Following is the view of V's in terms of issues and vulnerabilities:

- Volume: Tera-bytes of data is generated per day. To process this volume of data, a fine architecture is required. Data needs to be stored in forms of tables, files etc. This volume of data is tough to manage and process in distributed environments. An attack which can generate ambiguous data is tough to detect when mixed with normal data.
- Variety: Data obtained are structured, unstructured, single factor, multi factor, probabilistic, linked or dynamic in nature. Handling such variety of data is an issue for distributed database management systems. If an anomaly generates vague data with fast variation in data eats up resources with no fruitful output.
- Velocity: Data obtained in batch (group or cluster of data) or stream. Sometimes data processing requirements are real-time. Processing data incoming at such speed is an issue for processors. If an anomaly launches flooding attacks, it may be tough to detect, if flooding is done by mimicking the normal data.
- Veracity: Data changes its modeling from time to time. This needs to be handled every-time data changes its modeling. Therefore linkage is again an issue for such variability in data. An anomaly can implement it periodically with short duration of a period may again lead to resource wastage. It also includes the problem of trustworthiness, authenticity, origin, reputation, availability of resources and its accountability.

1.1 Security Issues in Big Data in Distributed Environment

Being Big Data widely used and adaptive technology, it is almost natural that immense security and privacy challenges arise frequently. Four V's of Big Data which are also characteristics of Big Data affects information security and give challenges to the design, functionality and management of architecture of big data. These security issues or challenges have a straight impact on modeling of security solutions that is necessary to handle characteristics and requirements of big data architecture in distributed environment. Figure 2 shows the various issues as security perspective in big data in distributed environment.

2 Infrastructure Based Security

Infrastructure of big data tools is a combination of multiple high performing computing clusters which are capable of processing huge data simultaneously.

These abilities comprise of management of data, cloud computing solution, and data analytics. Big Data Life cycle Management (BDLM) [8] model imposes suitable infrastructure in big data industries for implementation of the scientific method of data management. BDLM includes storing data in all stages of big data life-cycle that allows data re-usability. Here, we are considering software related security issues and not hardware related issues like hardware theft or hardware failure. Following are some key points regarding infrastructure based security:

2.1 Secure Computations

Let us consider Hadoop, a tool for big data processing which utilizes Map Reduce as the framework. In Map-Reduce [9], an input file is split into multiple small chunks. In the first phase, Mapper reads each chunk, performs computation and outputs a list of key-value pair. This is provided as input to next phase which is a reducer phase. Here, values belonging to each distinct key is combined and outputs the result. Usually, Hadoop utilizes map-reduce in distributed environment i.e. Master node gives the task to slave nodes and setup works as distributed system. So here are the security issues related to it:

- The untrusted node of the system could return a wrong result. This will, in turn, generate a wrong final output. Untrusted node points to the compromised system which is under the control of anomaly.
- With the large system, it is practically impossible to identify the error in mapper or reducer phase of Map-Reduce. it will take time to identify anomalous node. This may happen with other frameworks of big data processing tools as a majority of them processes data in distributed environment.

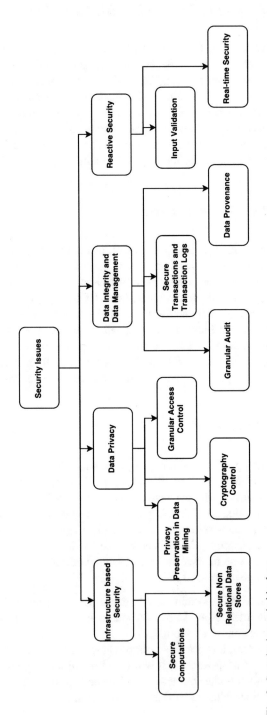

Fig. 2 Security issues in big data

2.2 Secure Non-relational Data Stores

NoSQL databases [10] are really popular to store non-relational databases. Since NoSQL is still newer than traditional databases like RDBMS, therefore it is natural that NoSQL is comparatively poor in terms of security as compared to the RDBMS [11, 27]. For example, there is no robust solution for NoSQL injection attacks [12]. It is well known that traditional database vulnerable to SQL injection attacks are prone to be used. NoSQL does not provide any built-in security mechanism [13] which can handle multiple attacks simultaneously. If we discuss in terms of NoSQL then following are the basic security issues:

- Industries are dealing with the problem of migration from relational database to non-relational databases which are still in development phase. It is not a good move to depend on No SQL completely.
- In general, databases like NoSQL depends on external enforcing mechanisms for security. Therefore industries who are trying to use Non-relational databases should be clear with the security policies of the third party implemented on the software.
- In distributed environments, vulnerabilities of non-relational databases increases as it is vulnerable to SQL Injections as well distributed denial of service attacks.

3 Data Privacy

Two terms closely related to each other comes into the picture: Data Privacy and Data Security. Data privacy refers to the suitably appropriate use of data. In distributed environment, if data is shared between two or more parties or systems then the data should be used to the agreed policies. Data security refers to the confidentiality, availability and integrity of the data in distributed environment. Data Security and Data Privacy goes hand to hand practically. Following are the two basic aspects:

3.1 Privacy Preservation in Data Mining

Privacy preservation [14] in data mining refers to the maintenance of privacy of a part of data such that no information can be gained from the anonymized part of data by the data miners. It is known that huge amount of information is hidden in the dataset. By proper data mining, these information can be extracted. Data mining can also become a security issue as it may reveal private information which needs to be hidden for personal, who is performing data mining). One of the live examples is AOL released search logs after anonymization [15]. But due to inefficient anonymization, miners were easily identifying the users in the logs. There are other security issues highlighted as:

- Data generated by industries, government agencies or institutes are usually continuously mine by analysts. A malicious insider or an untrusted partner can abuse these datasets and extract private information from data.
- When big data is passed to multiple nodes for processing (in distributed environment), if any of the nodes is under the influence of malicious user or contains a malicious code can easily result in providing data to the anomaly for mining.

3.2 Cryptography Control Mechanism

This is one of the critical security aspects for big data in distributed environment.

Since the data needs to be processed at multiple nodes, data needs to be passed through the network. Any anomaly can tap the network [16] or perform the man in the middle attack [17] to the grab the data. Therefore it is important to ensure that data must be secured from end to end and only readable to the parties whose sharing the keys. Specific research in the field of big data has been done in this field because usually big data processing is done in distributed environment. Attribute based Encryption [18] is an important research as it provides rich security, decent efficiency and scalability. Cryptography can give any data a high level of security but it cannot be applied as a whole on full data because of following reasons:

- Daily data stored remains unencrypted. The reason behind this is the computational complexity of cryptographic algorithms. Encryption and decryption is a highly time consuming process. It also consumes resources like computation power for encryption and decryption. As a solution, Attribute based Encryption is provided address this issue.
- Another reason is that the cryptographic algorithms either encrypt the whole data or none. Therefore if analyst needs to mine the data then it has all data purely visible or totally encrypted.

3.3 Granular Access Control

This basically maps to the secrecy of data. This implies that data should not be accessed by the people who do not have access to that data set. There are various softwares available to provide granular access controls. Suppose for a data set contains information about bank, school and hospital. If it is processed by big data processing tool containing one master node and three slave nodes then each node should have access to data of only one type i.e. of any one of bank, school and hospital. Any slave node should not have access to the data given for processing of other node.

- Big data analysis and cloud computing are increasingly focusing on handling diverse data sets. Handling such data set with the responsibility of maintaining the secrecy of varieties of data set during analysis and mining.

- Managing these restrictions on the dataset is a cost effective issue. Yet there are many software and tools available for maintaining granular access but still, they are costly.

4 Data Integrity and Data Management

Data integrity is one of the fundamental components of security [19] and proper data management is important to maintain data integrity. As a whole, it means the maintenance, assurance of accuracy and consistency of data throughout its life cycle. It is the critical aspect of design, implementation and usage of the big data processing systems, especially in the distributed environment.

4.1 Granular Audits

Granular Audit [20] basically points to the proper inspection of organization accounts to find improper behavior. Sometimes there are attacks which happened on the system cannot be detected by Intrusion detection system. Granular audit helps in getting information about those missed attacks. It helps not only in finding the reasons of missing attack detection but also helps in compliance and regulation of security design of the system.

- Login and Logout including the number of attempts in any machine can be found out. In the case of dictionary attacks, this is very useful.
- Commands issued to the system under administrative privileges from the same system or from the master node in the distributed environment can be verified from granular audits.
- Stored procedure executions of the systems.

4.2 Secure Transactions and Transaction Logs

A transaction is a unit of work performed within a database system. This system can be a standalone or in distributed environment. Mostly transaction occurs to make changes in the database. Change may be a new entry or the modification in the previous entry within the database. Secure transaction simply means the transaction happened from one end and the same transaction (without modification) has been transmitted to other end and reflected in the database as well.

- A transaction can be tapped by an anomaly. In a network, it may be in non-readable mode but the copy can be flooded to the destination. This may reflect the normal user who performed the transaction as an anomaly.

- Theft of transaction logs gives information about the transactor as well as the design of the database. It results in the privacy leakage and security breach.

4.3 Data Provenance

Data provenance refers to the origination of data. This term is defined in different ways in the literature. Provenance graphs are generated from provenance meta data. These are used to check the data dependencies. Analysis of such graphs some times results in the security or confidentiality leakage.

- Data provenance is a causality graph with annotations. The causality graph joins the participating objects which describe the process producing an object's current state. Each node depicts an object and edge between the nodes depicts a relationship between two nodes which are objects.
- It again reveals the indirect association of an object with another with attributes.

Anomaly may extract information from the graph.

5 Reactive Security

Security measures which are taken after the attack or breach happened once to avoid those type of attack in future come under Reactive security [21] measurements. Every organization takes some measures to prevent losses caused by anomalies. Each organization also plans to respond to such losses when the proactive measures either becomes ineffective or when they did not exist. Reactive methods of security include disaster management plans, use of investigation services and use of recovery specialists, re installation of operating systems and applications on compromised systems also known as zombies, or switching to alternate systems in other locations. Having an appropriate set of reactive responses prepared and ready to implement is just as important as having proactive measures in place.

5.1 Input Validation at Distributed Nodes

In various organizations and industries, big data collection happens at many end point devices. A key challenge in the data collection process is the input validation. A trust issue is always there with the collected data because it may contain malicious data or the artificial formed data which may result in inappropriate outputs. Input validation and filtering remain a challenge when data is collected from untrusted sources in the distributed environment.

- Suppose data is retrieved from weather sensor and also collected by the manual sources i.e. by using sensors of mobile phones and send the data manually may have a mismatch. An adversary can send the bogus data from a virtual sensor deployed in the same environment. Anomaly can spoof the phone ID to send the wrong temperature.
- To perform these type of evaluation proper algorithms need to be designed which may cluster out the anomalous data.

5.2 Real Time Security

Real time security monitoring [22] remains an issue in distributed environment.

This issue grows even bigger when it comes to Big data in distributed environment. Intrusion Detection Systems take time to process data and generates alarms but with latency. This is a high risk for big data processing. In general, false Alerts or alarms generated are inversely proportional to the time taken to process data for security. This means high false positives can be seen for Real time security tools. Following are some key points in terms of security:

- It is important to know about the entity using the data. It is also important to know about the access of that data to that entity in real time. This means the amount of time user is accessing the data. Entities online who are accessing data.
- Real time monitoring can help in early detection of attacks including worms, Zombies, Trojans etc.

6 Countermeasures

It is important not to just detect the attack and then try to overcome but also to prevent the attacks. There is work done in literature to keep big data processing in distributed environment safe. To keep environment secure, it is important to keep both system as well as network safe. Figure 3 shows a summarized overview of additional countermeasures with examples of products under that section.

Many of the measures are described in the above sections with the related security issues but few additional security measures to protect big data processing in distributed environment which are as follows:

- Anti-virus: Software which contains definitions of anomalous computer codes. In anti-virus, As soon as the malicious code is detected against the virus definition, an alarm is generated. Attackers when tries to insert anomalous code like virus, worms, zombies or Trojans in the systems, anti-virus detects it. Anti-virus also have capabilities to delete these anomalous executable codes. It is necessary to keep the definitions of anti-virus updated to have higher security in the system.

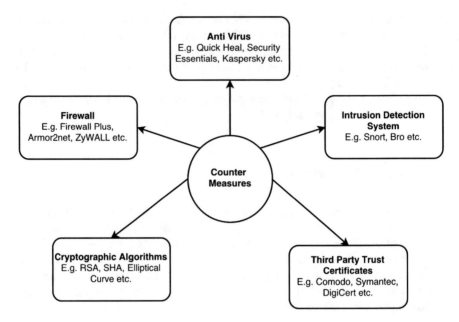

Fig. 3 Few counter measures

- Firewalls: It is a security software which monitors and controls the incoming and outgoing traffic. It allows or blocks the traffic on the basis of defined rules. Rules can be added, deleted or modified in the system as per requirement of the user under administrative privilege. Firewall usually blocks packets from untrusted sources and helps in system protection.
- Intrusion Detection Systems and Intrusion Prevention Systems: An intrusion detection system (IDS) is a type of security software designed to automatically alert administrators when someone or something is trying to compromise information system through malicious activities or through security policy violations. An intrusion prevention system (IPS) is a system that monitors a network for malicious activities such as security threats or policy violations. The main function of an IPS is to identify suspicious activity, and then log information, attempt to block the activity, and then finally to report it. Snort and Bro are famous IDS used in industries.
- Cryptographic Algorithms: It is always suggested to use those cryptographic algorithms which maintain the balance between processing time and cryptography key strength. Now a days elliptical curve cryptography is popular.
- Third party trust certificates: In this chapter, we have seen the importance of the third party especially for digital signatures. It should be a trust-able and known third party for maintaining policies between the two parties involved in the licensing.

These countermeasures can be used in different variants in the different type of security issues mentioned in the earlier sections. We cannot say that we can have one to one countermeasure mapping to security issues. It depends on the issue and how critical issue is to use the relative countermeasures. In addition to it, there may be other security solutions for providing secure big data processing in distributed environment. Therefore it is the sole responsibility of the administration to define the countermeasures for security related issues.

7   Conclusion

In this chapter, we had a bird's eye view of the security issues involved in Big Data Analytics applications executed in distributed environment. We have seen that it is important to protect not only the system but also the network in which distributed systems are placed. We saw possibilities of many attacks as it is a new technology and in its child state. We also gave some counter measures to these security issues. However, there may be different and difficult situations in which all these counter-measures may not succeed. Combinations of security software and tools can minimize the risk of attacks but as a base rule of security, issues can never reduce to zero.

It is our hope that, with the passage of time, more research will be done in the security aspects of big data in distributed environment which opens new doors to protect big data processing in distributed environment.

References

1. Chen M, Mao S, Liu Y et al (2014) Big Data: A survey. Mobile Networks and Applications 19(2): 171–209.
2. http://www.vcloudnews.com/every-day-big-data-statistics-2-5-quintillionbytes-of-data-created-daily/
3. https://www.sciencedaily.com/releases/2013/05/130522085217.htm
4. Jacobs A (2009) The Pathologies of Big Data. Communications of the ACM 52(8): 36–44.
5. White T (2012) Hadoop: The Definitive Guide. O'Reilly Media, Inc.
6. Shvachko K, Kuang H, Radia S et al (2010) The Hadoop Distributed File System. In IEEE 26th symposium on mass storage systems and technologies (MSST): 1–10.
7. Shanahan JG, Dai L (2015) Large Scale Distributed Data Science using Apache Spark. In Proceedings of the 21th ACM SIGKDD International Conference on Knowledge Discovery and Data Mining: 2323–2324
8. Li J, Tao F, Cheng Y, Zhao L et al (2015) Big Data in Product Lifecycle Management. The International Journal of Advanced Manufacturing Technology 81(14): 667–684.
9. McKenna A, Hanna M, Banks E et al (2010) The Genome Analysis Toolkit: A MapReduce Framework for Analyzing Next-Generation DNA Sequencing Data. Genome research 20(9): 1297–1303.
10. Han J, Haihong E, Le G et al (2011) Survey on NoSQL Database. 6th international conference on In Pervasive Computing and Applications (ICPCA): 363–366.
11. Leavitt N (2010) Will NoSQL databases live up to their promise?. Computer 43(2): 12–14.

12. Halfond WG, Viegas J, Orso A. et al (2006) A Classification of SQL-injection Attacks and Countermeasures. In Proceedings of the IEEE International Symposium on Secure Software Engineering: 13–15.
13. Okman L, Gal N, Gonen Y et al (2011) Security Issues in NoSQL Databases. In 10th IEEE International Conference on Trust, Security and Privacy in Computing and Communications: 541–547.
14. Agrawal R, Srikant R (2000) Privacy-Preserving Data Mining. In ACM Sigmod Record vol 29: 439–450.
15. Adar E, (2007) User 4xxxxx9: Anonymizing Query Logs. In Proceedings of Query Log Analysis Workshop, International Conference on World Wide Web.
16. Garfinkel S (2002) Network Forensics: Tapping the Internet. IEEE Internet Computing 6: 60–66.
17. Desmedt Y (2011) Man-in-the-Middele Attack. In Encyclopedia of Cryptography and Security Springer US: 759–759.
18. Goyal V, Pandey O, Sahai A et al (2006) Attribute-based Encryption for FineGrained Access Control of Encrypted Data. In Proceedings of the 13th ACM conference on Computer and communications security: 89–98
19. Hussain B (2006) U.S. Patent Application No. 11/425,524.
20. Futoransky A, Kargieman E, Bendersky D et al (2003). U.S. Patent Application No. 10/414,120.
21. Avramovic B, Fink LK (1992) Real-Time Reactive Security Monitoring. IEEE Transactions on Power Systems 7: 432–437.
22. Anthony E, Phillips J (2003) U.S. Patent Application No. 10/347,050
23. Tanenbaum AS, Vansteen M (2007) Distributed systems. Prentice-Hall.
24. Bertsekas DP, Tsitsiklis JN (1989) Parallel and distributed computation: Numerical Method vol 23.
25. Dean J, Ghemawat S (2008) MapReduce: Simplified Data Processing on Large Clusters. Communications of the ACM 51: 107–113.
26. Chaiken R, Jenkins B, Larson P et al (2008) SCOPE: Easy and Efficient Parallel Processing of Massive Data Sets. Proceedings of the VLDB Endowment vol 1: 1265–1276.
27. Ron A, Shulman-Peleg A, Puzanov A. (2016). Analysis and Mitigation of NoSQL Injections. IEEE Security & Privacy 2:30–39.

Scientific Computing and Big Data Analytics: Application in Climate Science

Subarna Bhattacharyya and Detelina Ivanova

1   Introduction

Analyzing and solving real world scientific and engineering problems are often computationally challenging. Understanding origins of the universe, Earth's weather and climate dynamics, enabling cancer drug discovery are some examples of such large-scale incredibly difficult problems in science and engineering. Intrinsically such problems are multi-dimensional, multivariate, nonlinear and non-stationary in their dynamics that do not have quick and easy closed-form computable mathematical solutions. Solutions to these problems involve complex mathematical modeling, simulation and analysis that are usually achieved by the use of highly sophisticated and expensive high performance computing [henceforth referred to as HPC] [1], using Super-Computers [2]. However, the advent of data-intensive science [3] has ushered a new era in the world of scientific computing, enabling scalable 'Big Data' [4] technologies and Big Data Analytics [5].

Current realm of science and engineering is getting redefined as we enter into an era of data-driven and data-intensive applications across all interdisciplinary fields from scientific discovery to business intelligence. The ease with which any and all information can be disseminated digitally in a cost efficient and scalable manner is phenomenal. Incredible opportunities are being created with the deployment of numerous sensors, advances in machine learning, artificial intelligence and visualization as new applications of data science open up. Such opportunities also face new technological barriers and challenges, due to the ever-increasing volume, velocity, and variety of information getting generated for mining and analysis.

S. Bhattacharyya (✉)
Climformatics, Fremont, CA 94555, USA
e-mail: subarna.bhattacharyya@climformatics.com

D. Ivanova
Scripps Institute of Oceanography, UCSD, San Diego, CA, USA

© Springer International Publishing AG 2017
S. Mazumder et al. (eds.), *Distributed Computing in Big Data Analytics*,
Scalable Computing and Communications, DOI 10.1007/978-3-319-59834-5_6

In this chapter we present an overview of how the problems associated with Scientific Computing can be addressed using Big Data Analytics. In particular, we focus on a real world complex scientific problem of Climate Change. Rest of this chapter is organized into three more sections. Section 2 broadly discusses the nature of Scientific Computing, their computational challenges, and two commonly used approaches, namely Super Computer based High Performance Computing and Cloud hosted Distributed Computing, to solve them. Next we discuss the challenges involved in Climate Analytics, as an example of Scientific Computing, in Sect. 3 and explore how Big Data Analytics can help to address them. In Sect. 4 we delve into details, showing how Earth's climate is modeled using complex fluid dynamics [6], and simulated, and how Big Data Analytics using Spark [7] platform enables processing of significantly large-sized output data in an integrated manner.

2 Computational Challenges in Solving Scientific Problems

In the last section, we briefly touch upon some examples of problems, solutions of which require extensive scientific computing. Such problems are often modeled using systems of simultaneous partial differential equations [8] that may be deterministic [9] or stochastic [10], homogeneous/inhomogeneous, with initial and boundary conditions. Usually there are no easy short-cut methods to solve these systems of equations. One often needs to resort to Monte Carlo [11] and finite element methods [12] to solve these systems of equations. For most part that process involves breaking down the large dimensions to infinitesimally small grid elements. Often that involves solving the complex dynamics for each such grid element, accounting for interactions between grid elements, and then assimilating and integrating results over millions of such grid points to obtain the bigger picture for solutions. Such simultaneous (parallel) computing capabilities over millions of grid points necessitate the hogging of large scale efficient compute resources. Such computations can be performed using high performance computing such as Super-Computers as well as using cheaper cost-efficient cloud computing alternatives.

'Super-Computers' are big monster machines built by companies like IBM, Cray etc. with extra-ordinarily high compute capabilities (Fig. 1) [13].

The compute capabilities of these machines are measured in terms of 'Petaflops' where 'flops' is an acronym for floating point operations per second. Petaflops is a quadrillion or a thousand teraflops or 10^{15} flops. For their extensive compute capabilities, supercomputers are expensive, with some machines costing upwards of $20 M each [14]. That makes their viability often very limited. Supercomputers are suited for large-scale highly-complex, real-time applications and simulations. Hence, traditionally, they are widely used in Scientific Computing which needs fast, iterative computations on large volume of data as well as complex interactive computation across large dimensions.

One problem that supercomputer users often face is the pain of handling and storing large size of output data. Usually as such output gets generated, they are

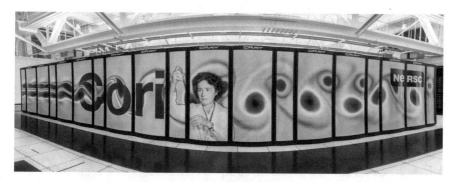

Fig. 1 National Energy Research Scientific Computing Center, is the frontier of high-performance computing sponsored by the U.S. Department of Energy's Office of Science. Located at Lawrence Berkeley National Laboratory its mission is to serve scientific research at national laboratories and universities. Shown is a panoramic view of the latest super-computer Cori – Cray XC40 with more than half million cores [13]

recursively stored away into HPSS (High Performance Storage System) storage or archives. Downloading this stored data in parts and analyzing them separately is often a very laborious exercise. More effort and time gets spent on effective managing of such data than in analyzing them to seek the exciting scientific findings.

In recent years, use of Distributed Computing frameworks for executing Scientific Computations is also slowly becoming a practice. The Scientific Computations that are not very sensitive to latency, can be reasonably handled by Distributed Computing frameworks. Especially Distributed Computing facilities provided by the Cloud vendors with on-demand access to a shared pool of configurable servers, storage, applications and services [15], is emerging as a viable alternative to run Scientific Computations on, at much cheaper cost and ease of accessibility. For instance, large volumes of pictures collected by Mars Rovers were processed on a computer cluster hosted on the Amazon Cloud by NASA [16] JPL efficiently, not hindered by small delays on individual computation. Similarly IBM's Cloud based Spark Platform is used by SETI Institute [17] in expanding its search for Extra-Terrestrial to include large-scale analysis of its 10-year archives (4.5 TB of data per hour), wide-band signal processing, and new long-duration observations.

Some recent studies have particularly focused on comparing the viability of using Distributed Computing frameworks in Cloud in detail over typical HPC using Supercomputers particularly in Scientific Computing [15, 18–20]. The open-source based software stacks in Cloud still poses some challenges for production science use [15], like dynamic scaling, multi-tenancy, standardization, reliability, security and privacy. However, these challenges can be addressed to reasonable extent from case to case basis depending on the type of Scientific Computation.

In the next sections, we shall discuss the challenges associated with Climate Analytics, an example of Scientific Computation, and the role of Big Data analytics in addressing it.

3   Climate Change and Big Data Analytics

Climate change is ubiquitous. It is affecting us all. With warming temperatures, increasing frequency of storms, floods, melting ice, rising sea levels, the first order impacts of climate change is often seen in large scale destruction left in the wake of unforeseen natural disasters. This in turn translates to economic and financial losses. Climate changes slowly and often creeps up in an unforeseen manner. For instance, one does not realize a drought at the onset of it, but likely to understand that when they are in middle of it. But there are second order impacts, for instance, cost to healthcare due to unforeseen climate-change driven diseases, cost to agriculture due to unknown plant diseases, or business interruption across various industry verticals. In fact a climate data company Weather Analytics, estimates that weather affects 33% of global GDP [21].

In order to address this challenging problem of understanding, and hence managing and thus adapting to changing climate, many countries in the world have come together to focus efforts to make a set of good action plan. Much of these actions need to rely on accurate, real-time or near-time predictive and prescriptive BigData analytics. There has been some progress towards that end in the development of tools such as Global Forest Watch, Microsoft Research's Madingley Model, and the Google Earth Engine, but still much remains to be done. The fast changing climate merits fast response in terms of generating and analyzing huge volumes of data to obtain accurate insights.

With enormous networks of sensors collecting data on all possible climate and weather variables like maximum and minimum temperatures, precipitation, humidity, soil moistures, solar irradiance, wind, and many more, Big Data analytics can help understand their interdependence and correlation that can then be used to come up with predictive and prescriptive models. This together with robust understanding of dynamics of weather and climate phenomena will be the tool that governments, businesses and stakeholders will need to use in order to mitigate impacts of climate change. In the next section we delve into the details of how we can model climate and then use it together with Big Data technology in an attempt towards predictive and hence prescriptive climate analytics.

4   Use Case on Climate Analytics

4.1   The Scientific Challenge of the Climate System

The Earth climate system is one of the most complex problems challenging the science today. It is multi-disciplinary, multi-scale problem complicated by the non-linearity of the interactions within the system. It consists of many components such as atmosphere, oceans, cryosphere, biosphere etc. which are interacting and changing in a wide range of time scales. The solar radiation impacts us every day. The fastest changing atmosphere controls the weather elements such as wind,

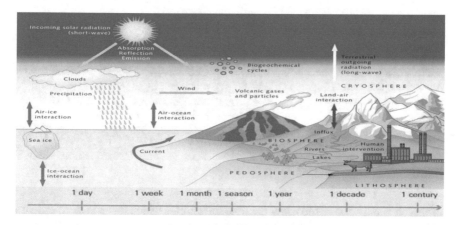

Fig. 2 Schematic representation of the major processes and components of the Earth's Climate System [29]

precipitation and clouds on a weekly basis. The more inert oceans are impacting the climate with phenomena such as El Nino on seasonal to interannual time scales. At least one new volcano is erupting every century and the large continental ice sheets and glaciers undergo changes over millennia (Fig. 2).

Centuries of scientific discoveries have synthesized the knowledge about each of the Earth's spheres (atmo-,aqua-,bio-,lito-,cryo-) in mathematical models including the primitive equations of fluid dynamics [22] describing the atmospheric and oceanic circulations, thermo- and hydro- dynamics describing the Earth's energy and water cycles, all based on fundamental physical principles of conserving mass, energy and momentum. These models are evolving continuously including new scales and processes. Most recently a new generation of Earth System Models has emerged in which bio-geo-chemistry processes have been added to the basic physical processes with the goal of accurate representation of the Earth's carbon cycle and explicit simulation of the green house effects.

The system of the fluid dynamics differential equations are solved by applying various numerical methods such as finite difference, finite elements transforming them into numerical equations discretized in space and time, requiring input of initial and boundary conditions. Initialized with the current climate state and with defined forcing conditions at the boundaries (e.g. the solar radiation at the top of the atmosphere, CO_2 forcing) the solution then will be a future state of the Earth system. This outlines the basic ideas of the climate and weather forecasting[1].

The complexity of the climate system is further convoluted by its inherent non-linear, chaotic nature [23] which makes it challenging to predict. Slight changes in the initial conditions or in the forcing at the boundaries will cause different trajectories in its time evolution. The difficulty is that the observed climate system time

---

[1]The weather and the climate are essentially the same phenomena but on different timescales. While the weather is the high-frequency component (from hours to weeks) the climate is the long term mean state of the climate system (30 years).

Fig. 3 Diagram of the
major components of the
Community Earth System
Model developed at
National Center for
Atmospheric Research
(http://www.cesm.ucar.edu/
about/)

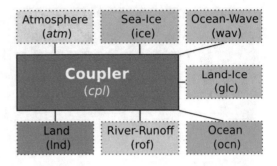

evolution is only one of the many possible trajectories which it can take. In order to
determine a range of possibilities or define solution uncertainty in our prediction we
need to run not one, but ensemble of many model simulations, which adds to the load
of computational needs and makes the problem computationally extensive as well.

Since the models are only an approximation of the real system they require continuous validation via comparison with the available observations. The historical
observational data sets date as early as the beginning of the twentieth century. The
earlier parts of the observational periods are often sparse and suffer from gaps in
time, largest over the oceans. During the satellite era in the last few decades we have
the observational data sets covering the entire globe with continuous high-frequency
time records. Nevertheless, the 30 years period of satellite record is too short to
evaluate the long-term fidelity of the model solution.

In the late 80s International Intergovernmental Panel for Climate Change (IPCC)
has been established to further our knowledge and understanding of the future of the
Earth's climate. The development of global observational networks together with
fast growing technology of high performing parallel computing facilitated further
the development of the climate models.

One of the most well-known climate models used worldwide is the Community
Earth System Model (CESM) developed at the US National Center for Atmospheric
Research (NCAR) in 1996 [24]. Figure 3 is a schematic representation of recent
version of the model which consists of atmospheric (atm), oceanic (ocn), ocean-
wave (wav), land (lnd), sea ice (ice), land ice (glc), river runoff (rof), interacting via
coupler (cpl) module.

4.2   Computational Challenge of the Climate Modeling

Solving the equations of a climate model requires compute power and therefore
through a variety of numerical methods the mathematical models are converted into
programmable algorithms in which the otherwise continuum of space and time of
the natural environment is discretized suitably in time and space grid

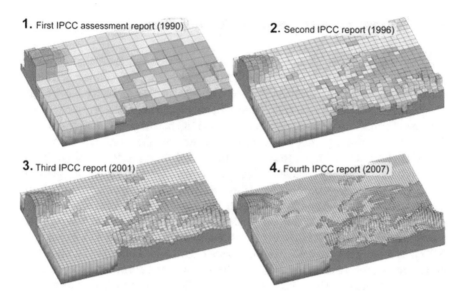

1. First IPCC assessment report (1990)

2. Second IPCC report (1996)

3. Third IPCC report (2001)

4. Fourth IPCC report (2007)

Fig. 4 Grid resolution evolvement of climate models used in the IPCC Assessment Reports

configurations. Inevitably, the accuracy of the numerical solution depends on the grid resolution. The smaller size of grid cells will result in smaller solution errors due to numerical approximation. The current generation climate models used to predict the future Earth climate by International Panel for Climate Change (IPCC) are commonly with 1 deg. (~100 km) grid resolution (or the smallest regions seen by this grid configurations are covering about 10,000 km^2 area) which is too coarse to resolve explicitly the fine scales of regional climate affecting our everyday life. Within the several decades of the Climate Model Intercomperison Project (CMIP) driven by the IPCC the climate models have increased their resolution 5 times, from ~500 km to ~100 km grid resolution, (Fig. 4, [25]). Phenomena like clouds and ocean eddies, which are sub-grid scale features, hinder the major source of uncertainty in the climate predictions. In order to build a climate model which includes explicit weather scale features like storms in the atmosphere and eddies in the ocean, we need numerical grid resolution of 10 km which scales computationally at a petascale level [26]. Going beyond this resolution sets a "Grand Challenge" for the currently existing computational resources.

An ongoing effort lead by the US Department of Energy is pioneering development of new generation of ultra-high resolution climate Earth system model [27, Accelerated Climate Modeling for Energy (ACME)] which will serve as a tool for better planning, decision and policy making of energy and water resources particularly in the era of climate change. The project roadmap will push the limits of the currently available high performance computing centers and it will utilize a new generation of exascale computer resources.

4.3   Post-processing Climate Model Output

The output that climate models generate are usually large 4 dimensional multivariate datasets varying from terabytes to petabytes in data size, the 'Climate Big Data'.
It is always a challenge to efficiently store, manage and analyze such huge data size
to extract meaningful science from them. Not only is the huge data size a challenge,
the usual format in which the data is stored and the language and platform needed
to analyze this data is often nonstandard. For instance the usual data files are in
netcdf [28] format and one needs to use the netcdf language to operate on them.
Although not commonly used in the industry, these languages, format and platform
are used by the climate modelers and other scientific communities around the world
and one can use python and R with netcdf libraries to operate these data files. The
large size of the data poses a resource problem for analysts because it is often not
possible to look at the data in its entirety because of insufficient memory resources.
Scientists usually have to download such data in parts which often is a laborious and
cumbersome process and then analyze each part and then repeat the process for all
such parts of interest. Then one has to piece together the analyses results and form
the bigger picture from them.

   With the advent of Big Data technologies like SPARK, this problem can be
addressed in a meaningful way.

4.4   BigData Climate Analytics Using Spark

The integrated platform of Spark not only offers a large memory in storage but
is also quite versatile in terms of programming paradigms (see Fig. 5). As shown
in this figure, the cloud infrastructure hosts the Spark platform and Data Storage.

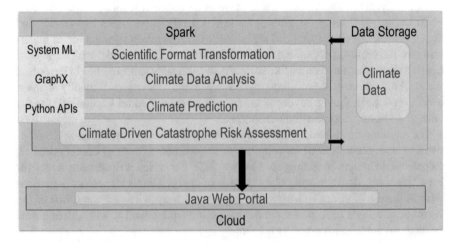

Fig. 5 A schematic diagram showing how Spark platform can be used to perform an end-to-end
climate analytics

The input climate data is stored in the Data Storage that is accessible to Spark platform. Spark platform supports programming in sql, python, scala and SparkR as well as analytical tools like Apache System ML (for machine learning), GraphX (for graph computing). These can be used to perform Scientific format transformation (from netcdf to csv formats), climate analytics, climate prediction and risk assessment. The output from these analytics are also stored in the Data Storage and can be accessed by web portal for visual display of output results. Thus such a platform can be used to ingest massive climate model datasets on which end-to-end climate analytics starting from data ingestion to customized climate prediction can be achieved.

We present some use-cases over select locations in California (see Figs. 6 and 7) to show how Climformatics (an early stage company in Climate Analytics) used SPARK for analyzing climate BigData towards customized climate prediction and

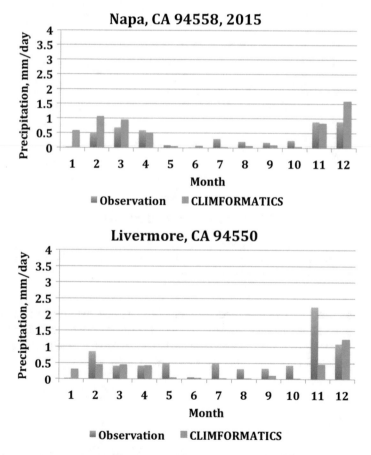

Fig. 6 Comparison of precipitation in mm/day for Northern California locations of Napa and Livermore in 2015. As predicted by Climformatics using Climate BigData Analytics on Spark (shown in *blue bars*) and observations (shown in *red bars*)

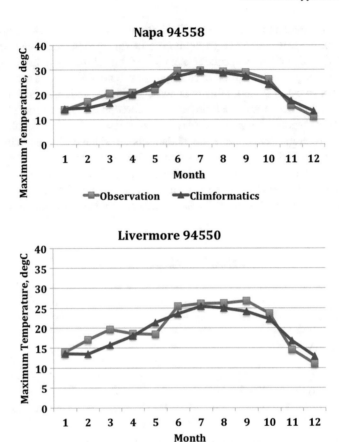

Fig. 7 Comparison of maximum monthly temperatures in degree Celsius for Northern California locations of Napa and Livermore in 2015. *Blue thick line* with marker shows Climformatics prediction, *red thick line* with markers shows observations

validate the accuracy of these hindcast predictions against historical observations. Such a tool can be used to obtain long-term climate driven catastrophe risk assessments and business insights particularly for businesses requiring large-scale multi-dimensional geospatial intelligence and data analytics.

5   Conclusions

Let us briefly recapitulate what we have discussed in this chapter and what are our main take-aways. We have learnt about the needs of scientific computing and briefly discussed about how those needs can be met by computational capabilities of HPC supercomputers or Cloud based Distributed Computing frameworks typically used in Big Data Analytics. As a use case, we looked at Climate Analytics problem. We then delved deeper into the challenges of climate modeling and prediction. Further we explored how climate modeling, analytics and prediction problem can be handled using Distributed Computing technology like Spark. In particular the Big Data technology platform like Spark, which uses Distributed Computing principles, enables scientists to take their compute analysis algorithms to the source of large sized storage data and execute such algorithms directly on the data without having to analyze it in parts due to limitations of memory and storage capacity. That opens up possibilities to do a lot more scientific enquiries and derive important insights from them much easily as compared to working the same using downloading data from HPSS storage in pieces and analyzing them.

With the rapid advances in Distributed Computing based Big Data technology and Cloud Computing services with the rapidly growing pool of enormous data information, there are growing synergies across different interdisciplinary sciences and engineering. The possibilities of doing amazing science and solving important problems facing the earth and its humanity, once believed to be intractable, seems to be within reach now through the use of Big Data Analytics.

References

1. https://www.techopedia.com/definition/4595/high-performance-computing-hpc.
2. https://en.wikipedia.org/wiki/Supercomputer.
3. Mork et al 2015, Contemporary Challenges for Data-Intensive Scientific Workflow Management Systems, Works 2015 2015 Austin, Texas USA.
4. https://en.wikipedia.org/wiki/Big_data.
5. https://www.sas.com/en_us/insights/analytics/big-data-analytics.html.
6. https://www.gfdl.noaa.gov/climate-modeling/.
7. http://spark.apache.org/.
8. Courant, R. & Hilbert, D. (1962), Methods of Mathematical Physics, II, New York: Wiley-Interscience.
9. S. Strogatz, Non-Linear Dynamics and Chaos: With Applications to Physics, Biology, Chemistry and Engineering (Perseus Books, New York, 2000).
10. Athanasios Papoulis and S. Unnikrishna Pillai, Probability, Random Variables and Stochastic Processes, 4th edition, McGraw Hill Boston, 2002.
11. Fishman, G. S. (1995). Monte Carlo: Concepts, Algorithms, and Applications. New York: Springer. ISBN 0-387-94527-X.
12. K. J. Bathe : Numerical methods in finite element analysis, Prentice-Hall Englewood Cliffs (1976).
13. http://www.nersc.gov/users/computational-systems/cori/configuration/.
14. http://insidehpc.com/hpc-basic-training/what-is-hpc/.
15. Sadashiv and Kumar 2011, Cluster, Grid and Cloud Computing: A Detailed Comparison, The 6th International Conference on Computer Science & Education (ICCSE 2011) August 3–5, 2011. SuperStar Virgo, Singapore.
16. https://www.datainnovation.org/2014/01/supercomputing-vs-distributed-computing-a-government-primer/.

17. https://www03.ibm.com/software/businesscasestudies/us/en/corp?synkey=P226318O092 27V08.
18. Roloff et al 2012, 2012 IEEE 4th International Conference on Cloud Computing Technology and Science.
19. Marathe et al, 2013, HPDC'13, June 17–21, 2013, New York, NY, USA.
20. The Magellan Report on Cloud Computing for Science U.S. Department of Energy Office of Advanced Scientific Computing Research (ASCR), December 2011.
21. http://www.kdnuggets.com/2015/12/big-data-predictive-analytics-climate-change.html.
22. Gill, Adrian. Atmosphere-Ocean Dynamics. (1982). Int. Geoph. Ser., Vol. 30, Academic Press
23. Lorenz, Edward 1963, Deterministic nonperiodic flow. J. Atmos. Sci., 20, 130–141.
24. http://www.cesm.ucar.edu/.
25. IPCC 4th Assessment Report, 2007.
26. Bader, David; Covey, Curt; Gutowski, William; Held, Isaac; Kunkel, Kenneth; Miller, Ronald; Tokmakian, Robin; and Zhang, Minghua, "Climate Models: An Assessment of Strengths and Limitations" (2008). US Department of Energy Publications. Paper 8. http://digitalcommons.unl.edu/usdoepub/8.
27. https://climatemodeling.science.energy.gov/projects/accelerated-climate-modeling-energy.
28. https://www.unidata.ucar.edu/software/netcdf/.
29. https://www.climatechangeinaustralia.gov.au/en/climate-campus/modelling-and-projections/climate-models/theory-and-physics/

Distributed Computing in Cognitive Analytics

Vishwanath Kamat

1 Introduction

Analytics is a method of logical analysis whereas Cognitive is involving conscious intellectual activity as thinking, reasoning, or remembering [1]. The science fiction writers have been describing super intelligent machines even centuries before Artificial Intelligence (AI) became discipline in academia. "The Turk", a hoax mechanical device that can play chess was invented by Hungarian illusionist Kempelen Farkas in 1789 [2]. Although *The Turk* was not an AI device, but it does represent human desire to augment their own intelligence. The mechanical arithmetic machine *Pascaline* developed by Blaise Pascal in early seventeenth century could be attributed to development of early computing devices and then further evolved into AI field of study as we know today.

The Cognitive Analytics Systems aka Cognitive Systems is a special type of Big Data Analytics systems. Like any other Big Data Analytics systems, Cognitive Systems have to learn from vast amounts of data, some times in the range of hundreds of millions documents, within reasonable timeframe. This needs processing of vast quantities of data in meaningful way to gather higher level of knowledge abstraction for a given dataset or field of study. Given the vast amount of data processing and computation required for such Cognitive Systems, use of distributed computing frameworks such as Hadoop, Spark, TensorFlow play vital role in building such system. In this chapter we shall delve into fundamental building blocks of cognitive Systems along with some key usecases.

V. Kamat (✉)
Lab Services, IBM Analytics, Dallas/Fort Worth Area, TX, USA
e-mail: vkamat@us.ibm.com

© Springer International Publishing AG 2017
S. Mazumder et al. (eds.), *Distributed Computing in Big Data Analytics*,
Scalable Computing and Communications, DOI 10.1007/978-3-319-59834-5_7

2 Building Blocks of Cognitive Analytic System

2.1 The Data Repositories

The Cognitive Systems's foundational feature is to learn from the data in an iterative fashion to identify trends and patterns to build analytical models. The effectiveness of such learning depends on type and quality of data that is provide during the learning period. Thus a Cognitive System should be able to ingest, manage and analyze variety of datasets often 100s of Terabyte scale.

Typically data stores used in cognitive system are based on variety and velocity of data assets that are being stored as input to the system. For structured data that needs to be queried and analyzed at large scale a system like Hadoop that include hive and hbase are fairly common. The Hadoop ecosystem works very well in distributed computing framework as underlying Hadoop filesystem stores data evenly across multiple systems. The execution frameworks such as Map/ Reduce and Spark within Hadoop ecosystem are data locality aware, so there is minimum network overhead while processing the queries in distributed fashion. For analysis that needs frequent random access to subset of data, traditional relational databases with MPP (massively parallel processing) architectures such as db2, Teradata and alike have been seen in use for decades. There are some variants of other databases such as MongoDB, CouchDatabase, Cloudant and other variants of open source databases are also good candidates as data stores in cognitive systems. The use of data repositories could be driven by type of data being collected such as JSON formatted data would be better handled by Couch Database where as key-value pair data for random access suited for Hbase repositories.

2.2 The Data Ingestion Tools

A data ingestion subsystem to load data as bulk or just-in time/real-time is one of the integral part of any cognitive system. The challenges faced in the data ingestion process often relates to the massive volume, variety and velocity of the data, often all three and sometimes combination of these three challenges. In this aspect Hadoop ecosystem of software fits the bill perfectly. The hadoop ecosystem software component such as Sqoop, Flume and Kafka can handle variety of sources including relational databases, flat file feeds, message queues and many more data types perfectly. For just in-time or real time processing Apache Spark (Spark Streaming) and Apache Storm are very cost effective frameworks that are very popular as well.

2.3 The Analytical Frameworks

The foundational capability of cognitive system includes software to analyze data and finding patterns and trends, building analytical models, training and scoring frameworks for models. Although basic algorithms are fairly similar for model building, the choice of analytical tools is often influenced by familiarity of the toolset among the data scientists working on the project, The advancement in computing power has enabled to crunch massive amounts of data and run complex algorithms against datasets in fraction of time it used take just a few years ago. The cognitive analysis, includes some of the most compute intensive processing such as NLP (natural language processing), ML (machine learning), probabilistic reasoning, and traditional statistical computation among other compute intensive data processing. The computation often requires the rapid prototyping and interactions to achieve desired outcome. Cognitive systems need to crunch huge amounts of data in order to learn the patterns and iterate through hypothesis. Often these cognitive systems are capable of learning on its own from new data as it churns through it without manual intervention.

The *machine learning* and *deep learning* are part of a broader Artificial Intelligence (AI) field of study. The *machine learning* sometime looked as a simpler version of *deep learning* that often includes multiple layers of Neural Network type of computation. The cognitive systems that use these deep learning techniques tries to gather higher level of knowledge abstraction for a given dataset or field of study. The cognitive system often uses "deep learning" algorithms such as Naïve Bayes, Decision Trees, and Neural Networks that include mathematical concepts such as "backpropagation". When the datasets are large, it is difficult to understand all the nuances of the data pattern for individual experts. There are few different type of machine learning algorithms such as "Supervised learning" and "Unsupervised learning" techniques used in this aspect of data exploration. Supervised learning consists of a defined target or outcome based prediction. The algorithms for Supervised learning rely on predictor variables or "independent variables" to derive target or "dependent variable". The computing algorithms include Regression, Decision Tree, Random Forest, Logistic Regression etc. Unsupervised learning consists of not knowing what to expect from the data and thus tries to find clusters of similar patterns or groups. Some popular use cases for such algorithm is used for customer segmentation using K-Means and Apriori algorithms. Each of these algorithms are compute intensive and often implemented to exploit parallel execution in a distributed fashion.

The machine learning and deep learning are key to any cognitive systems and often use a methodology called CRISP-DM (Cross Industry Standard Process for Data Mining). As shown in following diagram CRISP-DM methodology is an iterative process to build analytical model based on business requirements and understanding.

Cross Industry Standard Process for Data Mining (CRISP-DM) process [4]

As seen from CRISP-DM diagram, the analytical model development process revolves around data repositories and understanding of business relevance. It is often seen that data understanding and data preparation for modeling constitute 80% of the work involved in building analytical model. The evaluation feedback loop help maintain the currency of the model being developed and deployed in the long run. The analytical frameworks help in each steps of the CRISP-DM process by virtue of using sophisticated libraries for data cleaning, preparation, and often ease of deployment where constraints for production are quite different than development phase of model building.

The mathematical algorithms used in cognitive systems are implemented using various machine learning libraries such as R,Scikit-learn, nltk, Spark ML, Mahout and in recent days deep learning frameworks such as Theano, Caffe, DL4J, TensorFlow etc. These libraries and frameworks support various degree of computation efficiencies based on the problem it tries to solve. For example, Google's TensorFlow used by various google applications and gadgets for language translation and email analysis could be used with standard linux based system using CPU/GPUs as well specialized hardware called TPU (Tensor processing unit). The TPUs are similar in nature as CPUs but far less complex that are optimized to execute Tensor Flow, another machine learning framework efficiently.

2.4   The Hardware Components

The amount of data to crunch through to find patterns and trends is huge in typical cognitive systems. In order to sustain acceptable throughput for activities such as reading data and being able to analyze in quick iterations, massively parallel processing in distributed fashion is desired. The basic computing infrastructure for cognitive system typically will include several computers in a single cluster or set of clusters. These clusters of computers are built with low cost commodity hardware, each with massive IO, memory and CPU bandwidth. The Jeopardy winning Watson cognitive computer system was built with 90 computers servers with over 15 TB of memory and 2880 Power7 CPU cores [3]. The systems often use open source operating systems such as Linux and processing frameworks that are part of Hadoop ecosystems. In order for cognitive systems to be effective, these systems have to learn a subject or topic from several different sources often from variety of data sources including structured, unstructured and semi-structured datasets. There are various type of tools used to achieve such cognitive capabilities including machine learning, massively parallel processing (MPP) data repositories.

2.5   Key Non-functional Requirements to Consider

2.5.1   High Concurrency Throughput

The cognitive system could become integral part of a data processing pipeline where an action and response is chained across multiple systems. For example in a ATM cash withdrawal transaction, the network operator has to validate the card number and pin as basic check but a fraud detection system that is powered with cognitive processing capability may need to identify potential fraud within a second or two at max. In such use cases, the cognitive system needs to be able to sustain huge concurrent access often thousands of transaction per second. A distributed system such as large clusters of computers play a vital role in divide and concur paradigm for such cognitive systems. The cognitive systems while exploiting parallel infrastructure for process need to maintain consistency across whole infrastructure making it another challenge for syncing up distributed transactions.

2.5.2   Interfaces for Interaction with Systems

A consistent way to exchange information across different layers of the cognitive system is also a key requirement. This is often accomplished using REST (Representational State Transfer) API. The hadoop ecosystem enables REST APIs for various components for seamless exchange of information. The APIs help in abstracting the logic and processing of complex algorithms and enable it as service to end user applications. The APIs approach to such complexes computation help in avoiding to build and manage a large infrastructure required to train and maintain currency of the models while keeping the knowledge and expertise with the service providers, making it even better economic choice for end users.

2.5.3 High Availability and Disaster Recovery

The application of cognitive systems may result in being critical to businesses that a single point of failure could result in huge losses. For example failure of unavailability of systems with in an oil rig that could be monitored and controlled by cognitive system. In such application, the cognitive system needs to be made highly available i.e. should be configured with built-in redundancies to account for power failure, component failures, data center failure in case of disaster and be able to continue to function with minimal interruption. The systems in such scenarios need to account for making critical data and computing power available to the application seamlessly. This critical non-functional requirement often presents challenges to solution providers. The difficulty often stems from distance for disaster recovery site and backup/restore service level agreements within limited cost budgets.

2.5.4 Linear Scalability

One of the key feature of distributed computing is being able scale linearly when you add more hardware power, the system can handle additional workload without sacrificing performance. For example a cognitive system that is monitoring a network infrastructure for malicious attack needs to handle peak workloads during a special event such as launch of a product or handle seasonal increased demand during Christmas or holidays. The distributed computing framework fits perfectly for such systems which can divide and concur workload across dynamically allocated resources such as addition of a node to existing cluster infrastructure. The elastic nature of cluster computing provides the flexible framework to increase or decrease the amount of computing resources required based on workload presented to the system. The scalability features of the cognitive systems need to be designed not only for hardware scalability but also its software counter parts. The algorithms used by cognitive systems need to be designed to be able to exploit the distributed infrastructure.

2.5.5 Ability to Prioritize Workload

The cognitive systems when used in certain front end applications such as ATM and/or credit card transaction processing, needs to handle the priority of the requests that are flowing through the systems. For example, if during a peak demand for processing, the system should be able to bypass or halt a high dollar amount transaction over low value transaction for identifying fraud. The system with proper workload management, can prioritize the processing based on importance to the application being served. The systems such as ATM processing needs to handle the real-time and in cases where real-time response may be taxing the system, then it should be able to handle near real-time response seamlessly in order to minimize exposure to the fraud and misuse.

2.6   Cognitive System – Implementation Patterns

There are few technology patterns in cognitive system for implementation based on the use cases and certain functional and non-functional requirements. In its simplest form a cognitive system could be build using an on premise traditional cluster system that includes hardware, software and analytical tools built within a single homogenous system. For example a hadoop cluster deployed on premise using commodity hardware. The hadoop ecosystem of products includes analytical frameworks such as Spark and its components for building analytical models and execution pipelines. The hadoop ecosystem also includes products such as Hive and hbase for distributed data repositories. The open source community has created several frameworks to manage and maintain complex data structures such as Graph databases for network relationships among dataset being processed. There are cognitive systems that often need sub-second response to given event or action. In such cases, streaming frameworks such as Apache Spark Streams or Apache Flink become essential components of the ecosystem.

In recent days cloud patterns are maturing with different flavors as well. For example Tensor Flow offering in Google cloud, IBM Watson ecosystem offered in Bluemix (IBM cloud offering), Genereal Electric's Predix platform as a service geared mainly towards industrial applications, Amazon's cloud cognitive services in the form of select APIs for facial recognition, speech conversion etc. Some of these cloud offerings help developer to choose from simple API calls to sophisticated models that can be trained and maintained by end user applications.

3   Cognitive System – Use Cases

The cognitive systems have found their niche in various industries in recent years. The use cases described here showcases the breadth of the possibilities. The common theme across all use cases focus on machine learning, large scale data processing and easy interaction pattern with systems and end users. The use case in health care can be looked at any industry where body of knowledge is embedded in the associated literature and expert humans experience that can be utilized by wider audience which otherwise would have been impossible to imagine. The use case in Internet of Things could be viewed as any industry where large scale data gathering at various end-points is key and being able to act upon it intelligently and swiftly. The last use case around customer service will help understand how cognitive system could be used to build behavior patterns and predict possible proactive measures that can be taken to improve agility of any enterprise.

3.1 Cognitive Systems in Health Care

IBM's Watson Oncology is a cognitive system currently used by several doctors and healthcare providers across United States. The Memorial Sloan Kettering Cancer Center is using this system to individualize Cancer treatments to its patients. Although Watson Oncology system has been fed with various medical and research documents, the training of the algorithms was done by subject matter experts in the field. The system helps to correlate diagnosis with individual cases with vast research and early trial results to better serve patients. The Watson Oncology system is able to synthesize information in the massive quantity that will be near impossible for healthcare professionals to do in a consistent, repeatable manner. The system helps healthcare professional to identify new research drugs and trials that otherwise would have been limited.

One of the major benefits of cognitive healthcare system is that it is able to learn from new patient data and improve the analytical models it has already learned from historic information. The cognitive system provides interfaces to learn, connect and store vast amounts of data whereas the healthcare providers, researchers are becoming "trainers" to teach the analytical models by sharing their knowledge. When doctors ask questions in natural language, the Watson Oncology system is able to parse the patient conditions and show all treatment options with confidence level and effects so that doctors can choose best possible treatment option.

A typical flow of information in such a cognitive system will include a data repository or set of repositories to hold massive amounts of data. A special consideration needs to be given to the type of access that will be required such as batch vs real-time processing.

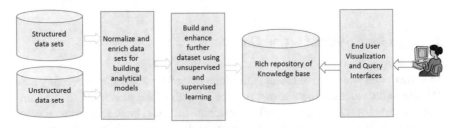

The above diagram shows a simpler view of information flow in a cognitive system focused on a particular subject area. The "normalize" and "machine learning" processes could be complex based on types of datasets involved. There is a significant training and review of the models required by subject matter experts to validate and tune the enrichment models deployed to build rich "knowledge base".

The hadoop ecosystem works very well in distributed computing framework as underlying hadoop filesystem stores data evenly across multiple systems. The execution frameworks such as Map/Reduce and Spark within hadoop ecosystem are data locality aware, so there is minimum network overhead while processing the queries in distributed fashion. For analysis that needs frequent random access to sub set of data, traditional relational databases with MPP (massively parallel processing) architectures such as db2, Teradata and alike have been seen in use. There are some variants of other databases such as MongoDB, Couch Database, Cloudant and other variants of open source databases are also good candidates for data stores. Each of these open source technologies have their pros and cons based on complexity and sometime non-functional requirements.

As discussed earlier for data ingestion, the system could employ real-time or near real-time framework based on technologies such as sqoop, flume, and Spark Streaming, Kafka etc. As part of data ingestion pipeline, it is possible to validate and enrich data as it arrives and augment the precision of the models to fine tune without human intervention.

It is key to identify patterns and trends within subject area being studied in a manner that the knowledge base is enriched with pre-determined clusters and/or associations among different attributes of datasets. The analytical framework like Apache Spark fulfills such requirements with help machine learning libraries it supports. The "unsupervised" models can derive such classification/association whereas the "supervised" models derive predictive behavior based on historical evidence as reference.

The visualization and interaction with the system could take various forms. The simple systems could be implemented using standard reporting and/or dashboard mechanism whereas the sophisticated system will need natural language processing (NLP).

To implement end to end a cognitive system, some of the key components come from systems such as Hadoop, various analytical libraries from Spark ML and R, repositories using Hive/Hbase or other noSQL databases. The system is then layered with API services for natural language processing and other API services to clean and enrich datasets. Each of these computational frameworks must exploit the parallelism in the distributed fashion in order achieve the scales needed for massive data repositories and running analytical workloads against such data sets to provide required service level objectives.

3.2   Cognitive Systems in Internet of Things Domain

The advent of smart phones and internet of Things (IoT) devices that are in use today has created new generation of threats. The security of Electric grid or drinking water supply in large cities have become vulnerable to such threats emerging from IoT devices. The e-meter change to electrical grid has provided enormous benefits to optimize the on-demand peak power generation. On the other hand, the e-meter devices, smarter substations, and distribution networks with billions of devices connected, opens possibility of attacks originating domestically as well as internationally.

The cognitive systems could take various forms such as improving consumer end points using smarter billing and usage to prevent theft and misuse to sophisticated systems that minimize outages and quick restoration of service after interruption. The complex web of millions of devices could be monitored and proactively maintain to reduce downtimes using predictive analysis of breakdown of components.

A cognitive system in such application will be able to handle massive amount of data at a very high frequency and be able to react in sub-seconds.

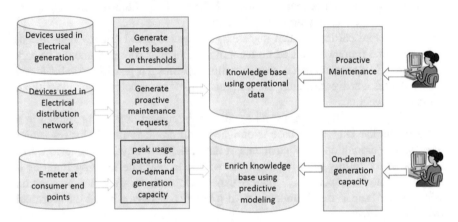

As seen in the above diagram, the data generated at power generation and distribution network is collected using sensors/IoT devices in a central repository. The type of data ingestion from such devices is often in real-time or in near real-time format. The real-time ingestion of data may pose some challenges as distance to central repository could be several miles apart. The mechanism such as "publish/subscribe" using Apache Kafka or other streaming protocol may be appropriate in such scenarios.

The data collected from various devices will be utilized to identify usage patterns as well as equipment functioning to gauge need for any proactive maintenance. The analytical models build to help with such scenarios may include linear regressions for anomaly detection or associations based on weather conditions and demand for power at a given point in time. As the data ingestion happens in real-time, it is possible to revalidate and fine tune the predictive model being utilized to generate alerts and feedback.

A cognitive system to cater to such functionality will include a data repository that can handle real time data ingestion, an inline predictive model scoring mechanism to provide instant probability of an incidence and feedback protocol, a alert and reporting capability for point in time or historic reporting. A distributed framework in such system allows the system to linearly scale as complexity of generation and distribution network increases.

3.3 Cognitive Analytics to Become a Customer Centric Organization

3.3.1 Next Best Action

The NBA (next best action) or BNA (best next action) refers to individualized marketing campaigns targeted based on unique characteristics of a customer. Unlike the marketing campaigns of yester years, where marketing or sales initiated programs that were offered as generic programs irrespective of weather likelihood of a customer utilizing is not considered. This type of campaigns were generally measured for success as aggregate conversions to sales often hitting low single digit success rates. With advent of new generation of computing frameworks and resources, it is possible to customize marketing offers based on individual customer behaviors pattern, demographics, past history and social activity. This type of campaigns has shown to produce significantly better success rate for the campaigns.

3.3.2 Changing Engagement Patterns

The generation of customers prior to millennium was primary engaged with business through physical postal mail, telemarketing and to some extent in recent year in the email form. The degree of engagement would have varied across different industries. The customers in the retail industry were targeted primarily using mail-in coupons and discounts whereas leisure and travel industries would have used telemarketing to sell their goods and services. The recent surge in use of social media fronts such as facebook, twitter, SMS, WhatsApp etc. not only from desktop computer but using handheld devices has created even more avenues for connecting with businesses. In order get a better understanding of customer it is becoming necessity to know how customer is connected in social media circles with others. It is becoming very important aspect for businesses to keep their presence in such social networking sites and domains.

The distributed technologies such as hadoop and open source documents database systems play vital role in capturing massive social interaction data.

3.3.3 360 ° View of Customer

In order better serve customers, it is required to know how customer is defined to business. It is possible that business may have different touch points to customer through sales, support channels as well as various subsidiary within parent business may interact with same customer. In order to provide accurate marketing campaign designed for individual customer, a single view of customer across different lines of business is required. To achieve such integration across multiple lines of business (LOB) a distributed computing framework becomes essential part of the system architectures required for integration.

As shown in figure below, it is essential to have common understanding of "who is the customer" to various lines of business is key. The LOBs could capture different attributes across businesses but will have value to each other for cross sell and upsell opportunities.

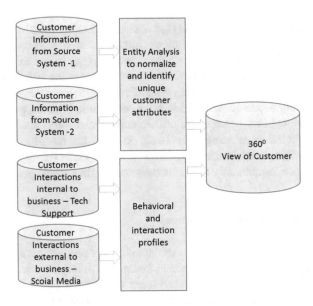

The repositories that store different records of the same customers could be stored in several flavors of RDBMS that needs to be extracted, transformed and loaded (ETL) into single repository to be made available to all LOBs. The process of ETL is another resource intensive that can exploit distributed computing frameworks. The "Entity Analytics" is another resource intensive process to identify unique attributes of each customer and network relationships across other customers. The tools in the marketplace such as IBM Information Server, Informatica and Abinitio are able to run parallel workloads in distributed fashion. There are specialized solutions for "Entity analytics" but open source does offer libraries and tools to build own solution as well. This process of consolidating customer records results in master customer record that then enable applications such as NBA and others could exploit.

3.3.4   Understand Thy Customer

The businesses today collect lot of data about customer and their interaction with business in various forms such as call center interaction over phone and online, website click-stream interaction, product specific review and discussion sites, popular apps such as "yelp" and "Consumer reports" etc. The businesses need to understand the impact of each of these interactions and build processes that can deliver overall a better experience with their business. For example if a customer is visiting a review site to research a product, then to influence the decision businesses can showcase their advertisement for their product om the same review site alongside the review and discussion board. The computational frameworks that can perform machine learning help to identify key patterns in the customer behavior from such interactions. To make this data about customer interactions useful, cognitive systems use the algorithms such as "Collaborative Filtering" and Content-based Filtering" that can derive patterns and recommendations within NBA applications.

Collaborative Filtering uses historical behavior of the customer and sometimes behavior of customers with similar characteristics to predict potential recommendations. Content-based Filtering uses historical behavior as well as content of the historical data that was used in the reaching such behavior. These types of algorithms often rely on explicit and implicit preferences of customer and description of products/services. The analytical model development and computation often done in a batch cycle to maintain behavioral history with recommendations that help NBA application to present just in-time marketing offer based on individual profile of customer.

The cognitive system that support NBA type application must process massive amount of data to consolidate various instances of customer data and merge to effectively represent it as true view to business with confidence. The behavioral patterns online and other means of interaction help build the rich profile that help in creating unique offers personalized to each customer. The distributed computing frameworks are able crunch such data and make it available for NBA.

4   Conclusion

As discussed before, it is obvious that the amount of data and computation needs to be carried out within short time span is enormous in order to find or guess patterns in Cognitive System. These patterns may look very similar across use cases but it may differ significantly in implementations specific to various industry contexts. Also, the identified patterns and its influence on future predictions may need to be revalidated on a frequent basis in order to maintain the accuracy of such cognitive systems. That also calls for substantial amount of data crunching within reasonable time frame. These patterns identified (and curated) would be eventually used as integral part of front-end as well as back-end system of enterprises in a seamless manner. All these necessitate the use of Distributed Computational frameworks and the technologies implementing the same (popularly called as Big Data Technologies) as the key building blocks of any cognitive system or application.

However, the implementations of various Distributed Computing constructs are taking various new generation approaches for better efficiency in terms of time and cost. Advent of Big Data Technologies like Hadoop, Spark, Flink, Kafka, etc., is making development of Cognitive Systems much more affordable (better cost/performance ratio) and fast compared to traditional technologies used in analytics.

The implementation of the Distributed Computing constructs in GPU based systems over standard CPU based systems for pattern identification and enrichment is also gaining huge momentum. Finally, use of these technologies either in on premise infrastructure or through Cloud based services or a hybrid of both (as Infrastructure, Platform or API as service) is providing numerous new avenues to integrate cognitive systems with other applications in commercial world.

References

1. "Cognitive." *Merriam-Webster.com*. Merriam-Webster, n.d. Web. 18 Feb. 2017.
2. B. Buchanan(2006)."A (Very) Brief History of Artificial Intelligence". AI Magazine Volume 26 Number 4 (20056) (© AAAI).
3. "Building Watson:An Overview of the DeepQA Project".(2010). Association for the Advancement of Artificial Intelligence. ISSN 0738–4602
4. Kenneth Jensen, 2013. CRISP-DM. retrieved from https://commons.wikimedia.org/wiki/File:CRISP-DM_Process_Diagram.png.

Distributed Computing in Social Media Analytics

Matthew Riemer

1 Introduction

These days Social Media Analytics is probably the most widely used Big Data Analytics application across various industries. In this chapter, we will discuss some of the most prominent use cases for Social Media analytics leveraged across the enterprise sector today. Social Media analytics includes strategies for leveraging intentionally public online interactions in order to drive business insights. This generally includes graph analytics to understand community networks, unstructured content analytics to understand shared content, and predictive analytics to drive supply chain optimization.

Specifically, we will address the use cases of influencer analytics and polling public opinion. Additionally, we will discuss using this analysis to forecast product demand. For each of these use cases, we will go into some of the most effective analytics techniques used to produce these insights. When done well, these capabilities can uncover significant benefits for organizations that use them. However, data and signal cleansing is a significant issue impeding business from properly leveraging social media data to generate real insights. Unfortunately, for this reason, only organizations that truly embrace cutting edge analytics techniques can find significant value in many cases.

For illustrative purposes to help readers understand the use cases presented in this chapter, we will periodically leverage prototype software developed at the IBM Watson Research Center. These and many other important capabilities have been significantly enriched as part of the IBM product Watson Analytics for Social Media.

In this chapter, we will proceed by first discussing commonly used open source software for social media analytics, and then go into detail about best practice tech-

M. Riemer (✉)
AI Foundations Lab, IBM T.J. Watson Research Center, New York, NY, USA
e-mail: mdriemer@us.ibm.com

© Springer International Publishing AG 2017
S. Mazumder et al. (eds.), *Distributed Computing in Big Data Analytics*,
Scalable Computing and Communications, DOI 10.1007/978-3-319-59834-5_8

niques influencer analytics and social polling use cases. Finally, we will conclude by discussing using these insights for predictive modeling.

2 Open Source Tools for Social Media Analytics

Because quality and efficient analytics is of premium concern for Social Media Analytics use cases, Apache Spark is a natural starting point for a Big Data platform. Apache Flume and Apache Kafka are very popularly used for ingestion. For analyzing unstructured text data, it is easiest to use software for indexing. This greatly reduces the computational overhead of searching for some content over potentially many billions of user interactions. Popular alternatives for indexing are Apache Solr and Elastic Search. Elastic Search is especially good to consider if intensive geo-spatial analytics is required. Moreover, structured data is commonly stored in HDFS and accessed with Spark SQL.

Apache Spark provides the GraphX library for efficient graph processing that is very useful in influencer analytics and community detection use cases. Some Machine Learning algorithms are also available in an optimized form through Spark MLlib. However, for access to the most advanced and cutting edge Deep Learning capabilities it would be easiest to leverage another open source framework like Caffe or Tensorflow that has proven out Apache Spark integration.

3 Influencer Analytics

The rise of social media has brought with it a lot of fascination among the public with the idea of social influencers. These social influencers are critically important to the success of various social media applications as they keep people coming back. They can also influence the behavior of social media users by providing products with exposure. As such, proper utilization of social influencers has become a key component of modern branding and advertising.

3.1 Understanding the Impact of Influencers

In order to understand more about the role influencers have in the emerging social landscape, many research studies have been devoted to understanding their impact. This is particularly true for Twitter where interactions are largely publicly accessible. For example, Cha et al. [1] carried out a comprehensive study on influencers on Twitter. They considered in-degree (the number of connections the user has), number of retweets, and the number of mentions as parameters for assessing the influence of a user. In their study of the influence of about 6 million users on a population of 54 million users they arrived at a few findings of interest. First, the in-degree of a node is not necessarily and indicator of top influence. They additionally found that top influencers are often an influencer for not one but multiple topics. Moreover, they found that influence is not built over night, rather with a long term concerted effort. Bakshy et al. [2] looked at the cascade of tiny URLs in order to gain insight about Twitter influence. They explored 74 million cascades of tiny URLs among a population of 1.6 million users. Their conclusion was that longer cascades were mostly started by influential users. Other papers have focused on variants of the PageRank algorithm to understand influence. A scalable version of PageRank is available as part of Apache Spark in its GraphX Library. A good example is "Twitterrank" [3] which has been shown to produce results that are very different from and sometimes qualitatively better than methods that focus on counting in-degree, retweets, and mentions.

Chiefly, there are two aspects of social media relevant for analysis of influencers:

Graph Methods First, is the follower graph and diffusion of posts over it. These social graphs can be analyzed by using popular graph analytics techniques like PageRank that are often based on the notion of centrality. A very notable graph analytics library for use at Big Data scale is GraphX of the Apache Spark project. It is argued in Embar et al. [4] that it makes sense to use not one, but multiple interpretable graph based metrics to assess the influence of a user. In that work they show influence can be quantified by user graph centrality, social media activity, the response rate to their posts, the response volume to their posts, and the number of followers they have.

Content Methods Secondly, it is important to look at the content of posts. In influencer analytics it is rare to see use of sophisticated text analytics machine learning techniques. This is because a lot of important information can be ascertained by simple analysis of post with logical rules. Generally, hashtags are used to understand key topics and other topics are filled in by setting topic monitors. Topic monitors generally include a list of terms to search through for each post in order to flag relevant conversations. There have been some efforts to leverage popular unsupervised machine learning techniques like Word2Vec [5] to help users expand terms, which makes this kind of analysis even easier [6].

3.2 Wimbledon Influencer Case Study

We will now follow a real-world example of influencer analytics. Figure 1 shows an analysis of the influence of Roger Federer on Twitter at Wimbledon during a year in which he was heralded as the "Wimbledon Twitter Champion" by the London Evening Standard.

Engagement refers to the volume of responses to his tweets. Activity is the number of messages he posted during the time period of interest. Authority is a measure of if he is connected with or his content is shared with other authoritative people. Timeliness refers to the response rate to his tweets. Followers is just a count of his

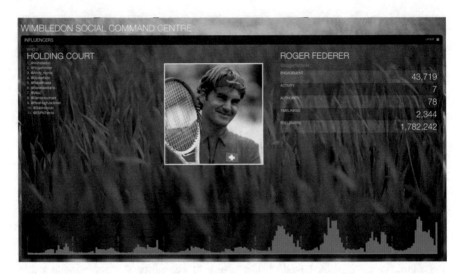

Fig. 1 An analysis of Roger Federer's influence on Twitter during Wimbledon leveraging five different metrics

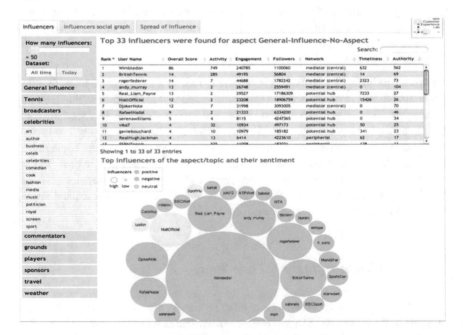

Fig. 2 An analysis of influencers at Wimbledon and their sentiment across all topics

Rank	User Name	Overall Score	Activity	Engagement	Followers	Network	Timeliness	Authority
1	Wimbledon	86	3	1590	1098369	mediator (central)	6	805
2	ESPNTennis	60	55	1046	181860	mediator (central)	37	500
3	ESPN3	22	27	345	138789	mediator (central)	0	204
5	Faitelson_ESPN	13	0	219	1077965	mediator (central)	89	136
6	fersch_espn	12	0	203	362159	mediator (central)	0	126
4	espn	19	1	365	9916595	mediator (central)	0	111
7	Telegraph	11	0	209	521118	mediator (central)	0	100
9	BBCSport	10	5	160	2345774	mediator (central)	6	76
8	TwitterSports	11	0	222	3307815	mediator (central)	0	71

Fig. 3 An analysis across influence metrics of different media organizations on Twitter at Wimbledon

twitter followers. Figure 2 looks at Roger Federer's influencer at Wimbledon relative to other influential Twitter user handles. It also considers topics of discussion at Wimbledon and sentiment. Other tennis players that had a large degree of social influence at Wimbledon that year include Andy Murray, Novak Djokovic, Rafael Nadal, and Serena Williams.

The next chart, Fig. 3, looks at different media organizations and their influence on day 1 of Wimbledon that year.

ESPNTennis was the most authoritative on Twitter for the first day. ESPNTennis and ESPN in general had the highest influence. They were more impactful than BBCSports, Telegraph, Guardian, and so on. This is interesting because ESPNTennis has 180,163 followers to the 2,326,280 followers of BBCSports.

4   Social Polling

Another very popular use case for Social Media Analytics is social polling. Social polling solutions are motivated by the thought that in contrast to the time and money spent on polling ideas within a small focus group of people, you can consider a much wider net of people that voluntarily express opinions about some topic.

4.1   Sentiment Analysis

A key capability for this use case is sentiment analysis, which is determining the positive, negative, or neutral feelings of a speaker. Logical human derived rule based approaches have been proposed for this use case, which mostly rely on positive/negative sentiment word counting and negation analysis. The obvious downside of these approaches is that they need to be manually made from scratch. It may be difficult to build the term lists needed by hand without missing many relevant words.

In this case, semi-automated term suggestion tools based on unsupervised machine learning can help a lot [6].

Another very popular approach is using supervised machine learning techniques on annotated social media data. Supervised machine learning techniques are generally initialized without any prior knowledge and conduct all of their learning on the labelled training data provided. Supervised Deep Learning techniques like Recurrent Neural Networks [7, 8], or Convolutional Neural Networks [9] are increasingly becoming the method of choice for these use cases based on strong empirical results. However, bag of words approaches that treat each word as a unique element in a fixed vector that ignores word order are still very popular in industry. Generally, bag of words representations are used as features for Naïve Bayes, Maximum Entropy, SVM, or Random Forest machine learning models.

Deep Learning is a field of machine learning focused on the application of originally biologically inspired neural network models. Neural networks rose to popularity in the machine learning community in the 1980's and early 1990's, and were largely abandoned before recently achieving breakthrough results across many machine learning problems. The rise of Deep Learning models over the last five years has significantly advanced the state of the art in sentiment analysis. At the same time, significant quantities of training data are generally needed to achieve great results. Annotating training data can be expensive and it is difficult to select a training set that is fully representative of testing conditions. For these reasons, we generally only have comparatively little labelled training data available. Unfortunately, "narrow AI" that knows nothing other than what it is taught in the training data performs poorly in this setting. To achieve good results without an ideal amount of training data, the most competitive machine learning social media sentiment analysis models leverage prior knowledge external to the dataset. For example, the winner of the SemEval 2016 shared task, Swiss Cheese [10], leverages 90 million external tweets with noisy labels. This leverages a popular strategy of using massive amounts of tweets that contain emoticons in order to understand associations between words and sentiment at a large scale which was first innovated by [11]. Another very popular strategy is unsupervised feature sharing. This is when unsupervised features extracted from massive amounts of unlabeled data are used as base features for a classifier on top. Popular examples of this are Word2Vec [5] and Glove [12] that have been shown to build high quality unsupervised representations of words. These representations can also be built on the sentence level [13] or document level [14]. An alternative strategy with similar motivation is unsupervised pre-training. This strategy differs from feature sharing in that the shared representation is further tuned on the labelled training data as recently shown very effective for sentiment analysis in [15]. This kind of strategy is good when you have enough labelled target task data to improve the representation.

Another important family of techniques are those including Multi-task Learning, Knowledge Transfer, and Lifelong Learning that take knowledge of one task to improve their ability to do another. For example, [16] the current state of the art Twitter sentiment analysis technique leverages knowledge from a Common Crawl of the internet, Movie Reviews, Emoticons, and a human defined rule logic model to drastically improve the performance of its recurrent neural network model.

Two advanced topics involved in sentiment analysis are aspect oriented sentiment and domain adaption. These are tough problems as they involve many considerations to be done well. Aspect oriented sentiment does not produce a document level sentiment, but sentiment analysis specific to each item where there is sentiment in the document. Many aspect oriented sentiment analysis systems use a dependency parser as a pre-processing step to make it easier to focus the system. Not doing so generally requires a significant amount of training data in order to do the end to end task effectively. Domain adaption is adjusting a classifier built in one domain to another domain. Popular techniques in the research community leverage Grassmannian manifolds and only unlabeled in domain data in order to learn domain invariant representations for the original labelled task. The learned models tend to perform significantly better than those with no adaption. A recently proposed Deep Learning technique that tackles the problem of domain adaption is Domain Adversarial networks. Domain adaption is a very hot research topic that is likely to make major strides in the next ten years as the community moves away from "narrow AI" and towards general purpose AI models.

4.2 Intent Detection

One very important social media analytics problem is intent detection. This refers to when someone explicitly expresses that they are going to buy a product or go to an event. To create a machine learning method capable of general purpose intent detection has been considered an NLP challenge problem with high confidence. As such, in industry we have seen logical rule based intent detection systems become very popular. One example of a popular system is IBM SystemT [23] that has the ability to match logical rules with granularity based on a semantic parsing of a sentence. The parser allows for higher accuracy rules matched to the linguistic patterns related to extracting intent. Intent detection is a process best done today by product domain experts to pick up on domain specific terminology and lingo for expressing intent.

4.3 Topic Monitoring

In order to make social polling possible, a sentiment analytics solution needs to be tied with topics of discussion. This sentiment about topics is the core insight of social polling that provides client value. There are two major paradigms for monitoring topics in social media posts which we will refer to as "top down" and "bottom up".

Top down topic monitoring is when someone wants to explicitly search social media for trends about a topic. This is how the Google search engine works. The most common approach to this problem in social media is the use of query engines such as Elastic Search that facilitate matching against lists of terms combined via Boolean logic. The Boolean OR operator is used to compile terms together that form a sub-topic. The Boolean AND operator is used to create a "must contain" condition for multiple sub-topics within a social media post. The Boolean NOT operator is used to create an exclusion rule often used for commonly confused topics that are unrelated. Unsupervised term expansion techniques can be really helpful in creating lists joined by the OR operator [6]. However, some recent Deep Learning techniques have explored a more ambitious solution to this problem, focusing on building document representations that are adequate for topic search, including [13, 14, 17].

Another example of top down topic monitoring is image analytics in Social Media. The main idea is that some posts contain most of their meaning in images. These solutions are top down because you generally need to build an image classifier ahead of time for the specific thing you want to search for. Recently and especially since [18], a deep learning technique called Convolutional Neural Networks has become the method of choice overtaking traditional approaches that rely on feature engineering. A big catalyst for this has been the development of the massive ImageNet dataset that contains millions of images comprising 1000 categories. One common deployment scenario is using a pre-trained model, such as those available in the Caffe model zoo, that already gets superior results on ImageNet. ImageNet categories can be a good starting point for image topic analysis. Other techniques require potentially fine-tuning a model developed on ImageNet with new labels or using the ImageNet model's hidden features as a basis for constructing a higher level reasoning model for new data. Common use cases include developing classifiers for images of specific products or influencers. Large social media companies such as Facebook build their own Deep Learning classifiers to classify images as specific users.

Bottom up topic monitoring is different from top down monitoring in that it leverages the content of the tweets itself in order to aggregate important topics without explicitly searching for them. Deep Learning techniques have recently shown the ability to perform high quality assessments of the semantic relatedness between two pieces of text for example in [19–22]. It has also been demonstrated that with N^2 (where N is the number of posts to search over) similarity computations, high quality cohesive topics can be derived. However, when N can be over a billion posts included in a search, this computation becomes computationally infeasible with reasonable current hardware. As such, approaches like [17] that combine using a machine learning based semantic similarity computation with extremely fast keyword match style search to focus on a smaller group of posts can lead to much more manageable computation in a bottom up fashion.

The most popular form of bottom up topic monitoring is information extraction. Information extraction pipelines such as those included in IBM SystemT [23] and

GATE's Annie pipeline [24] generally come loaded with a full host of capabilities starting with very low level analysis like part of speech tagging and word group chunking. Higher level capabilities include Named Entity Recognition [25], and proposed for Twitter in [26]. Named Entity Recognition identifies words that are included in the same entity, and the type of entity that it is. In systems made for the news domain like the CoNNL 2003 shared task, the entity types are usually Person, Location, Organization, and Miscellaneous. However, in the social media domain we are usually interested in even more detail. For example, the WNUT 2016 Twitter Named Entity Recognition shared task dataset includes categories like Product, Geo Location, Facility, Person, Company, and Other. Perhaps the most difficult element of a typical information extraction pipeline to produce good results for are parsing tasks like Dependency Parsing, and Constituency Parsing. There have been at least initial experiments [27] applying parsing to the Twitter domain, but good datasets are needed to make further progress. Parsing attempts to totally deconstruct the meaning of text in a pre-defined way that can be used for down stream analysis.

Once topics have been monitored at the tweet level, aggregations are generally made to understand the current amount of tweets about a topic, the velocity of the growth of each topic, and the acceleration of each topic. Topic acceleration is a principled way to understand which topics may be rapidly growing whispers and more in depth approaches have also been considered as in [28].

4.4   User Segmentation

Beyond what people are saying and when people are saying it, it is important to think about who is saying it. By this we rarely mean identifying exactly who said it, but more so what kind of person said it. Demographic analysis is a key aspect of understanding the projected impact of a few individual's being randomly sampled for polling on an entire population. As such, it is vital component of polling both for elections and market research. Techniques for dividing users into categories vary in difficulty based on the level of high level inference needed for the task and the availability of user contributed metadata for a particular social media platform.

Some social networks, like Facebook, are largely successful in getting people to submit structured data about their demographics and interests. However, the quality of user submitted data varies significantly across platforms. Twitter, for example, has much lower reliability and less information included in a typical profile bio. When profile attributes submitted to the user are reliable, they can be the easiest way to segment users. Sites that have large quantities of reliable self-reported data have a huge advantage in analyzing user micro-segments. The next easiest way to segment users is when they directly mention or express sentiment towards a topic that is explicitly being monitored for. This can be monitored pretty easily by applying logical matching logic to a list of topic terms. This type of query of easily accessible with Big Data ready text matching software like SystemT [23], ElasticSearch, or Solr.

The most difficult types of user segmentation require deep inference and higher level reasoning about the contents of a user's post. Good examples of this include inference of age and inference of income. It is generally very difficult and costly to create training data for a setting like this one, so it is important to leverage machine learning techniques utilizing prior knowledge to get good results. For example, in [29] large scale unsupervised learning and transfer learning were used in combination to predict the age range of twitter users based on the contents of their tweets.

User segmentation based on post content is hard and generally not possible without collecting a large history of posts for each user. This makes deployment of these solutions difficult if you are using a pay for query service. This is because it is likely pivotal for good inference to have access to posts about a user that are not related to the query topic. As such, it is significantly more efficient for data providers and their partners to infer these attributes about users ahead of time and provide it along side the received query data. There are multiple offerings in industry that provide these kinds capabilities for purchase, allowing for more useful social polling insights [30, 31].

4.5 Some Social Polling Examples

The following real-world social polling solution example in Fig. 4 leverages IBM software used to understand insights about customer segments by geography and topic category.

Additionally, it is useful to consider topic trends and customer segments associated with them. As you can see here, customer segments can be much more specific than the broad demographic categories used in traditional election polling as you

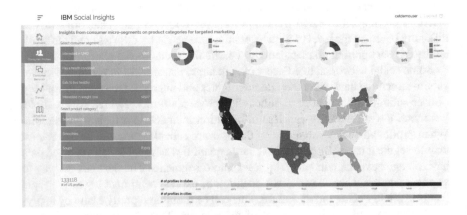

Fig. 4 An analysis of customer segments by geography and topic category

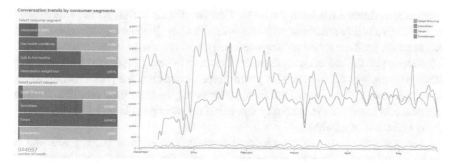

Fig. 5 An analysis of conversation tends over time by customer segment

can keep track of stated interests and hobbies for more refined insights. Social polling is a very good way to keep your organization up to date about societal trends that effect your business the most and their prominence within different categories of your customers. As an example, in Fig. 5, we demonstrate an analysis of the trends of monitored topics across different categories over time.

4.6 Social Polling for Demand Planning

In many industries, demand forecasting techniques have hit a wall in performance over the last decade. As popular univariate techniques like Holt-Winters and ARIMA models have staid stable for some time, the opportunity for additional gains in performance now will seemingly rely on taking this analysis a step further by incorporating external data to the time series for modeling. This can potentially explain volatility in the signal that we traditionally considered to be "random anomalies" that we will only attempt to smooth out when using most univariate forecasting techniques.

Recent work has applied Deep Learning to multi-factor forecasting based on many factors external to the forecast [32]. The success of these techniques to learn good representations without extensive feature engineering by human analysts opens the doors for possibilities in modeling external trends to a forecast in order to explain currently unexplained forecast volatility. Common low hanging fruit factors to consider in these models include weather and price, which could create economic value through significantly improved forecasts for organizations prepared to use them. Trends present in social media present a significant opportunity to explain human behavior that may effect consumer trends. Riemer et al. [32] found social media trends to be nearly as influential to retail demand forecasting a week ahead of time as weather. Indeed, applying social media trends to time series problems has become an emerging trend in the industry, especially with the rise of Twitter [33, 34].

One of the most straightforward ways to use social media to help predict product sales is too look for mentions directly related to the product and its competitors on social media. Measuring the intensity of conversation over time and with analysis of sentiment and intent has been shown to allow for rich forecasting data in attempting to explain volatility not show with univariate models. In general, the same style of information used for social polling use cases should be useful in this setting as well. Unfortunately, the effects are generally shown to be most useful for forecasting in the near term. Indeed, to create long term forecasts that include social media trends, it often includes predicting social media trends over the same time span. This can be difficult and is unlikely to account for some of the most prominent trends that seemingly come out of nowhere.

Another good use case for applying social trends to time series prediction is augmenting popularity information about events. For example, some events like Christmas have an effect far before their actual date. We can use social mentions as a proxy for understanding the natural ramp up and ramp down of an event. It is also useful for quantification of how big an event is likely to be this year. For example, we can project since everyone is talking about the Oscars early than they did last year and at a stronger intensity that it is likely to be a big Oscars this year. We can query systems built to predict based on external factors to understand how much of an impact, for example, uncertainty about the popularity of the Super Bowl can have on expected demand for various products.

An important aspect of demand forecasting based on external factor solutions is that they need to be interpretable to human analysts. This is especially true when they are predicting a big change in the forecast. This is a primary reason why deep learning attention models have been shown well suited to multi-factor forecasting [33]. Another emerging trend in Deep Learning is the great success of Reinforcement Learning since the Deep Q Networks idea proposed in Nature [35]. OpenAI has recently released "Universe" a massive open dataset meant to help push research forward in this area [36]. It makes more sense to think of forecasting as the reinforcement learning problem of planning stock. In actuality, there are many complex requirements about planning that are disconnected from the prediction of demand based on previous sales and more aligned with the actual profitability of a business. Looking forward, we should see the emergence of better techniques for this problem as the research community continues to push forward with Deep Learning and Reinforcement Learning.

5 Conclusion

In this chapter, we have gone over some successful analytical choices for effectively executing influencer analytics, social polling, and product demand prediction based on public social media interactions. Our focus has been on graph based, machine learning based, and logical rule based analytical strategies for achieving these use cases. To enable these use cases, it is best to build off a Big Data platform that embraces efficient analytics like Apache Spark, Solr, and Elastic Search. The limiting factor for what can be achieved with these solutions today is in many ways mostly technological. Analyzing unstructured data is very difficult and doing so can be very manual, which increases the time to value for adding new analytical facets. Organizations that have had the most success have strongly followed new techniques for unstructured analytics. In particular, this includes the huge achievements of Deep Learning in recent years for Natural Language Processing and Computer Vision. Social Media Analytics has the potential to see a large increase in capabilities in the years to come as cutting edge machine learning techniques advance our ability to do analytics on unstructured data. A particularly interesting recent trend is the rise of machine learning for doing classification based on a very small number of examples [37–41]. As these techniques advance to the point where they can be reliable for text classification and image classification tasks, we can see once again a big decrease in time to value for creating new facets of capabilities that should drive even bigger benefits for social media analytics based use cases.

References

1. Cha, M., Haddadi, H., Benevenuto, F. and Gummadi, P.K., 2010. Measuring user influence in twitter: The million follower fallacy. *ICWSM, 10*(10–17), p. 30.
2. Bakshy, E., Hofman, J.M., Mason, W.A. and Watts, D.J., 2011, February. Everyone's an influencer: quantifying influence on twitter. In *Proceedings of the fourth ACM international conference on Web search and data mining* (pp. 65–74). ACM.
3. Weng, J., Lim, E.P., Jiang, J. and He, Q., 2010, February. Twitterrank: finding topic-sensitive influential twitterers. In *Proceedings of the third ACM international conference on Web search and data mining* (pp. 261–270). ACM.
4. Embar, V.R., Bhattacharya, I., Pandit, V. and Vaculin, R., 2015, August. Online topic-based social influence analysis for the wimbledon championships. In *Proceedings of the 21th ACM SIGKDD International Conference on Knowledge Discovery and Data Mining* (pp. 1759–1768). ACM.
5. Mikolov, T., Sutskever, I., Chen, K., Corrado, G.S. and Dean, J., 2013. Distributed representations of words and phrases and their compositionality. In *Advances in neural information processing systems* (pp. 3111–3119).
6. Heath, F.F., Hull, R., Khabiri, E., Riemer, M., Sukaviriya, N. and Vaculín, R., 2015, June. Alexandria: extensible framework for rapid exploration of social media. In *Big Data (BigData Congress), 2015 IEEE International Congress on* (pp. 483–490). IEEE.
7. Hochreiter, S. and Schmidhuber, J., 1997. Long short-term memory. *Neural computation, 9*(8), pp. 1735–1780.
8. Cho, K., van Merriënboer, B., Bahdanau, D. and Bengio, Y., 2014. On the Properties of Neural Machine Translation: Encoder–Decoder Approaches. *Syntax, Semantics and Structure in Statistical Translation*, p. 103.
9. Dos Santos, C.N. and Gatti, M., 2014, August. Deep Convolutional Neural Networks for Sentiment Analysis of Short Texts. In *COLING* (pp. 69–78).
10. Deriu, J., Gonzenbach, M., Uzdilli, F., Lucchi, A., De Luca, V. and Jaggi, M., 2016. SwissCheese at SemEval-2016 Task 4: Sentiment classification using an ensemble of convolutional neural networks with distant supervision. *Proceedings of SemEval*, pp. 1124–1128.
11. Go, A., Bhayani, R. and Huang, L., 2009. Twitter sentiment classification using distant supervision. *CS224N Project Report, Stanford, 1*(12).
12. Pennington, J., Socher, R. and Manning, C.D., 2014, October. Glove: Global Vectors for Word Representation. In *EMNLP* (Vol. 14, pp. 1532–1543).

13. Kiros, R., Zhu, Y., Salakhutdinov, R.R., Zemel, R., Urtasun, R., Torralba, A. and Fidler, S., 2015. Skip-thought vectors. In *Advances in neural information processing systems* (pp. 3294–3302).
14. Le, Q.V. and Mikolov, T., 2014, June. Distributed Representations of Sentences and Documents. In *ICML* (Vol. 14, pp. 1188–1196).
15. Dai, A.M. and Le, Q.V., 2015. Semi-supervised sequence learning. In *Advances in Neural Information Processing Systems* (pp. 3079–3087).
16. Riemer, M., Khabiri, E., and Goodwin, R., 2016. Representation Stability as a Regularizer for Improved Text Analytics Transfer Learning. *arXiv preprint arXiv:1704.03617*.
17. Mitra, B., Nalisnick, E., Craswell, N. and Caruana, R., 2016. A Dual Embedding Space Model for Document Ranking. *arXiv preprint arXiv:1602.01137*.
18. Krizhevsky, A., Sutskever, I. and Hinton, G.E., 2012. Imagenet classification with deep convolutional neural networks. In *Advances in neural information processing systems* (pp. 1097–1105).
19. Tai, K.S., Socher, R. and Manning, C.D., 2015. Improved semantic representations from tree-structured long short-term memory networks. *arXiv preprint arXiv:1503.00075*.
20. Bowman, S.R., Angeli, G., Potts, C. and Manning, C.D., 2015. A large annotated corpus for learning natural language inference. *In Empirical Methods in Natural Language Processing (EMNLP) 2015*.
21. Rocktäschel, T., Grefenstette, E., Hermann, K.M., Kočiský, T. and Blunsom, P., 2015. Reasoning about entailment with neural attention. *arXiv preprint arXiv:1509.06664*.
22. Cheng, J., Dong, L. and Lapata, M., 2016. Long short-term memory-networks for machine reading. *arXiv preprint arXiv:1601.06733*.
23. Krishnamurthy, R., Li, Y., Raghavan, S., Reiss, F., Vaithyanathan, S. and Zhu, H., 2009. SystemT: a system for declarative information extraction. *ACM SIGMOD Record*, *37*(4), pp. 7–13.
24. Bontcheva, K., Derczynski, L., Funk, A., Greenwood, M.A., Maynard, D. and Aswani, N., 2013, September. TwitIE: An Open-Source Information Extraction Pipeline for Microblog Text. In *RANLP* (pp. 83–90).
25. Tjong Kim Sang, E.F. and De Meulder, F., 2003, May. Introduction to the CoNLL-2003 shared task: Language-independent named entity recognition. In *Proceedings of the seventh conference on Natural language learning at HLT-NAACL 2003-Volume 4* (pp. 142–147). Association for Computational Linguistics.
26. Ritter, A., Clark, S. and Etzioni, O., 2011, July. Named entity recognition in tweets: an experimental study. In *Proceedings of the Conference on Empirical Methods in Natural Language Processing* (pp. 1524–1534). Association for Computational Linguistics.
27. Kong, L., Schneider, N., Swayamdipta, S., Bhatia, A., Dyer, C. and Smith, N.A., 2014. A dependency parser for tweets. In *Proceedings of the 2014 Conference on Empirical Methods in Natural Language Processing (EMNLP)*.
28. Mathioudakis, M. and Koudas, N., 2010, June. Twittermonitor: trend detection over the twitter stream. In *Proceedings of the 2010 ACM SIGMOD International Conference on Management of data* (pp. 1155–1158). ACM.
29. Riemer, M., Krasikov, S. and Srinivasan, H., 2015. A deep learning and knowledge transfer based architecture for social media user characteristic determination. *SocialNLP 2015@ NAACL*, p. 39.
30. https://console.ng.bluemix.net/catalog/services/insights-for-twitter
31. http://datasift.com/
32. Riemer, M., Vempaty, A., Calmon, F.P., Heath III, F.F., Hull, R. and Khabiri, E., 2016. Correcting Forecasts with Multifactor Neural Attention. In *Proceedings of The 33rd International Conference on Machine Learning* (pp. 3010–3019).
33. Chen, Z. and Du, X., 2013, September. Study of stock prediction based on social network. In *Social Computing (SocialCom), 2013 International Conference on* (pp. 913–916). IEEE.

34. Nguyen, L.T., Wu, P., Chan, W., Peng, W. and Zhang, Y., 2012, August. Predicting collective sentiment dynamics from time-series social media. In *Proceedings of the first international workshop on issues of sentiment discovery and opinion mining* (p. 6). ACM.
35. Mnih, V., Kavukcuoglu, K., Silver, D., Rusu, A.A., Veness, J., Bellemare, M.G., Graves, A., Riedmiller, M., Fidjeland, A.K., Ostrovski, G. and Petersen, S., 2015. Human-level control through deep reinforcement learning. *Nature*, *518*(7540), pp. 529–533.
36. https://openai.com/blog/universe/
37. Lake, B.M., Salakhutdinov, R. and Tenenbaum, J.B., 2015. Human-level concept learning through probabilistic program induction. *Science*, *350*(6266), pp. 1332–1338.
38. Santoro, A., Bartunov, S., Botvinick, M., Wierstra, D. and Lillicrap, T., 2016. Meta-learning with memory-augmented neural networks. In *Proceedings of The 33rd International Conference on Machine Learning* (pp. 1842–1850).
39. Vinyals, O., Blundell, C., Lillicrap, T. and Wierstra, D., 2016. Matching networks for one shot learning. In *Advances in Neural Information Processing Systems* (pp. 3630–3638).
40. Kaiser, L., Nachum, O., Roy, A. and Bengio, S., 2017. Learning to Remember Rare Events. *In ICLR 2017.*
41. Kaiser, L., Nachum, O., Roy, A. and Bengio, S., 2017. Optimization as a model for few shot learning. *In ICLR 2017.*

Utilizing Big Data Analytics for Automatic Building of Language-agnostic Semantic Knowledge Bases

Khalifeh AlJadda, Mohammed Korayem, and Trey Grainger

1 Introduction

In the era of Big Data Analytics, search and recommendation engines have become the primary mechanisms through which users both actively find and passively discover useful information. As such, it has never been more critical for these data systems to be able to deliver targeted, relevant results that fully match a user's intent.

Search and recommendation engines can barely compete unless they leverage models containing deep insights into the kinds of questions being asked and - more importantly - the kinds of answers being sought. One of the most common ways of representing a domain in order to surface these insights is through the use of ontologies - combinations of taxonomies containing known entities, their properties, and their interrelationships. These ontologies can then be integrated into a search application in order to improve its ability to meet the end-user's information need. For example, if someone searches for the term *server* in the information technology domain, it has a very different meaning (a computer server) than in the restaurant domain (a waiter/waitress), and if someone is using a job search engine, this query could actually represent either meaning depending upon the user's context.

Ontologies are usually built manually by human experts, making them expensive to create, maintain, and update. To combat this, ontology learning systems, which attempt to automatically learn relationships from a domain and then map them into an ontology, are becoming more prevalent [1].

K. AlJadda (✉) • M. Korayem • T. Grainger
CareerBuilder, Norcross, GA, USA
e-mail: khalifeh.aljadda@careerbuilder.com; mohammed.korayem@careerbuilder.com;
trey.grainger@careerbuilder.com

© Springer International Publishing AG 2017   137
S. Mazumder et al. (eds.), *Distributed Computing in Big Data Analytics*,
Scalable Computing and Communications, DOI 10.1007/978-3-319-59834-5_9

In this chapter we will discuss techniques and algorithms that utilize the power of big data analytics and distributed computing to automatically build languageagnostic semantic knowledge bases. Such semantic knowledge bases enable significantly better query and document understanding and, as a result, drive much more relevant results to any given search or recommendations query.

We will cover some of the core technologies that enable such a system to be built (Apache Lucene/Solr, and Apache Hadoop), and will walk through some practical details of how such a semantic search engine has been built and is being leveraged in a real-world implementation.

2   Search Engines

Search engines are one of the most common ways through which people interact with digital information, and they can benefit tremendously from the integration of semantic knowledge bases, which improve the search engine's overall ability to accurately interpret and respond to queries. The underlying data structures within the search engine, as we'll later see, are also ideal for auto-generating and modeling those same semantic knowledge bases. Within the field of information retrieval, search engines are the tool of choice for enabling adhoc querying of free-text information (typically keywords) within massive amounts of content (up to trillions of documents), while simultaneously ranking and sorting the results by their relevancy to the incoming query. Most of the time, search engines are expected to do all of this work in milliseconds, or at most seconds.

This ability to search for any combination of keywords across up to trillions of documents and rank the relevancy of all the results to the query with sub-second response times requires some purpose-built data structures and data modeling approaches running in parallel across a distributed system. Chief among these are an inverted index, sharding and replication of data, a denormalized datamodel, and a distributed aggregation and scoring model.

2.1   Key Technologies

The search engine technology we utilize in our real-world example is Apache Solr, the popular open source search server [2]. The Apache Lucene/Solr project is the world's most popular open source search project, with most search engines today being built on top of Lucene and/or Solr. The rest of this section will describe how search engines, such as Apache Solr, achieve their large-scale distributed search capabilities.

2.2   Inverted Index

An *inverted index* is the mechanism by which fast keyword lookups are made possible. While the underlying implementation of an inverted index can be very sophisticated in order to optimize the speed of lookups and maximize the compression of data to fit as much as possible into memory, the basic structure is very straightforward. To build an inverted index, it can be useful (though not necessary) to first build a *forward index* which maps each document to the list of terms contained within the document. This is useful for looking up a document to see which words it contains, but it is less useful if you are trying to find which documents match a given set of keywords, as you would have to loop the list of words for every document to determine if the any queried keyword was found within the document.

Instead, search engines rely on an inverted version of this index, which maps each keyword to the set of documents which contain it, for $O(log\ n)$ time lookup of any keyword. An example of how a set of documents would be represented in both a forward index and an inverted index can be found in Fig. 1.

One piece of information potentially lost in a simplistic inversion of the forward index would be the position of the terms within the document, so these term positions (often along with additional meta data) are stored along with the document identifier in what's called a *postings list* associated with each term in the inverted index.

Whenever a query is executed against the inverted index, a lookup is performed on the inverted index for each term in the query, and set operations can be performed on the sets of documents matching each term to quickly resolve arbitrarily complex Boolean queries (e.g. *nurse AND hospital, java OR scala*). Phrase queries (i.e. *"brown fox"*) can even be resolved by leveraging the positions within the postings list to filter down to documents where all of the terms occur in sequential positions.

2.3   Sharding of Data

One of the additional beneficial characteristics of the inverted index is that, as the number of documents indexed increases, the probability of the terms from those documents already existing in the index also increases. This means that, with large scale data sets, the number of terms in the index will plateau, while the number of documents can continue to increase, since the number of terms is mostly a function of the distribution of terms within the underlying language(s) of the documents.

This makes it possible to easily partition an inverted index into multiple subindexes and to subsequently distribute queries to each of the indexes in parallel and simply aggregate the resulting documents. This partitioning of the index is often referred to as *sharding* the index. This parallel searching and aggregation across shards can be done across a multiple networked computers, enabling search engines to search across billions or even trillions of documents in well under a second.

Documents

id: 1
job_title: Registered Nurse
desc: a registered nurse at hospital doing hard work
skills: oncology, phlebotemy

id: 2
job_title: Software Engineer
desc: software engineer at a great company
skills: .net, c#, java

id: 3
job_title: Java Developer
desc: a software engineer or a java engineer doing work
skills: java, scala, hibernate

Forward Index

field	doc	term
	1	a
		at
		doing
		hard
		hospital
		nurse
		registered
		work
	2	a
		at
desc		company
		engineer
		great
		software
	3	a
		doing
		engineer
		java
		or
		software
		work
job_title	1	Registered Nurse
...	...	...

Inverted Index

field	term	postings list	
		doc	pos
		2	4
	a	1	1
		3	1,5
	at	2	3
		1	4
	company	2	6
	doing	1	6
		3	8
	engineer	2	2
		3	3, 7
desc	great	2	5
	hard	1	7
	hospital	1	5
	java	3	6
	nurse	1	3
	or	3	4
	registered	1	2
	software	2	1
		3	2
	work	1	10
		3	9
job_title	java developer	3	1
...	...	...	...

Fig. 1 Mapping documents into a forward index and inverted index

2.4 Replication of Data

Just as sharding makes it possible to increase the speed of queries across enormous numbers of documents and to scale beyond the capacity of a single server, it is often also necessary for a search engine to be able to handle large numbers of queries at a time. When the capacity of a single node to handle the volume of incoming queries to a shard of the index is exceeded, a replicated copy - or *replica* - of that shard can be placed on another servers such that the incoming queries can be load balanced across each of the replicas.

One additional benefit of replicas is that they can be used to provide fault tolerance within the search cluster. Since servers will fail from time to time, if at least one replicated copy of each shard exists on a separate server, then the search cluster can continue successfully responding to queries with no data loss as long as they re-route requests for that server to another replica of that shard.

2.5   Denormalized Data Model

The ability to shard an inverted index, create replicas of those shards, and distributed queries and indexing across a distributed cluster of servers enables search engines to scale in almost any direction (faster response times, more data, more queries). A critical data modeling rule must be followed to enable this parallelization, however - adhering to the use of a denormalized data model. In traditional relational database management systems (RDBMS), the best practice is to normalize tables and join on foreign keys as relationships between multiple tables in order to prevent data redundancy and inconsistencies from arising within the database. While this works well in theory, it prevents one from being able to easily shard out an index, since the requirement to join on separate indexes means you must have those indexes fully present on each server so that you can efficiently perform the join. While a few modern search engines (such as Apache Solr) do support join functionality, it must be used in a very intentional and cautious way in order to preserve the scaling characteristics of the search cluster as well as ensure all joins have access to the correct data such that they resolve correctly. Treating every document as a full representation of all of its denormalized, related fields is the recommended approach for building large-scale distributed search engines.

2.6   Distributed Aggregation and Scoring

One of the most important features of a search engine is the ability to score the relevancy of each document to the query matching it, and to return all matching documents in a sorted order. That sorted order is usually the calculated relevancy score, but the sorting could also be based upon the value of any other field or function.

In order for a distributed search engine to be as efficient as possible, it needs to maximize the work done in parallel on each shard, while minimizing the number or network requests and the amount of data being transmitted in order to arrive at a final, sorted result set to return.

While each search engine calculates relevance scores slightly differently, most use statistics easily derived from the structure of the inverted index. Calculations leveraging tf-idf (term frequency * inverse document frequency) scores, such as the popular BM25 scoring algorithm, consider the number of times each term in the query appears within each document (the *term frequency* or *tf*) multiplied by how significant that word is believed to be to the query (the *inverse document frequency* or *idf*). Term frequency can be calculated by the number of positions a term occupies within a document in the postings list, whereas inverse document frequency can be calculated leveraging the size of the set of documents in the postings list for the term. While getting a perfectly accurate *idf* score across all shards would technically require each shard sharing the *idf* of each term in the query (which isn't that expensive), due to the nature of word distributions within language, each shard in a randomly partitioned index will often have approximately the same document frequency distributions for each term, allowing relevancy calculations to be done completely independently on each shard.

Thus, in order to return a final list of relevancy ranked results, at a minimum a query just needs to be distributed in parallel to each shard of an index, each shard must then independently lookup the set of documents matching each keyword and perform the appropriate set intersections based upon the query, the resulting documents must then be sorted using a relevancy score calculated from statistics available in the inverted index for that shard, and then a set of results large enough to satisfy the requested number of documents must be returned to the aggregating node within the cluster. The aggregating node then just needs to re-sort the final returned documents from each of the distributed shards and then filter that list to the number of documents to return to the end user.

There are many additional features such as faceting/analytics, highlighting, and spell correction available in most search engines which may add additional distributed steps to this workflow, but fundamentally they all operate in this parallel way across the shards to enable sub-second query execution across billions or trillions of documents.

One last particularly important characteristic of search engines is that, as users issue queries and see results, they interact with those results. They may click them, skip over them, or even issue additional queries to correct their own spelling or try additional related keywords to see if they will yield a more relevant result. In the later Semantic Discovery section, we will describe performing large-scale data mining of this log data as a key technique for automatically building semantic knowledge bases.

3 Recommendation Systems

Recommendation Systems (RSs) automate the process of discovering the interests of a user by utilizing knowledge discovery and data mining techniques in order to predict items of interest to individual users and subsequently suggesting what should be relevant to his/her needs [3, 4]. Over the years, techniques and applications of RSs have evolved in both academia and industry (e-commerce/e-shopping, e-library, e-learning, e-tourism, etc.) due to the exponential increase in the volume of data. RSs can be broadly clustered into three main categories: Content-Based (CB) [5, 6], Collaborative Filtering (CF) [5, 7], and hybrid techniques [8]. Contentbased recommendation systems are the most sensitive of these to understanding the textual content since these RSs rely on matching items/users based on the similarity between their textual description. Thus, the availability of a semantic knowledge base is crucial for improving the performance of content-based recommendation systems [9].

While recommendation systems are often built as stand-alone systems which can match content to users' interests, they also overlap heavily with the functionality of a search engine. Whereas search engines are typically thought of as accepting a query and returning results matching that query, search engines can also leverage information about users and their preferences to personalize the search results. Likewise, whereas recommendation systems are often thought of as utilizing behavior of users to recommend based upon their tastes, it is very often helpful to be able to adjust recommendation matching in real-time and to be able to perform matching based upon arbitrary content and features, which is a task performed exceedingly well by search engines. Indeed, many modern recommendation system implementations are delivered through an inverted index, including the employment search use case we will be highlighting, enabling real-time recommendations and adjustment of features and their scoring coefficients for matching dynamically through simple query manipulation versus requiring code changes. Both kinds of systems (search engines and recommendation engines) can be thought of as existing along a fluid continuum of relevancy-driven information retrieval engines.

For these kind of relevancy-driven information retrieval engines to function optimally, it is important that they have some grounding in the domain they are providing information retrieval for, versus just being a bunch of generic algorithms. As such, being able to access some kind of semantic knowledge base which represents an understanding of the concepts and relationships within the domain is critical to them adequately performing the task of meeting users' information needs.

4 Semantic Discovery

Building semantic knowledge bases has traditionally focused on utilizing ontologies/taxonomies which are manually built and maintained, or employing clustering and dimensionality reduction to discover latent semantic links among terms of a given corpus. Building manual ontologies/taxonomies is not scalable, is hard to maintain, and is very labor expensive. On the other hand, dimensionality reduction is prone to noise and is not entirely human-understandable. Instead, we rely on search logs which turn out to be a rich source for discovering semantic relationships between phrases. In this section we discuss how to use distributed big data analytics for mining search logs to discover semantic relationships between key phrases in a manner that is language-agnostic, easy to interpret, scalable since it utilizes the power of distributed computing, and mostly accurate. The specific implementation of our technique will be presented in the context of an employment search engine in English, but the technique is both domain- and language-agnostic.

4.1 Problem Description

To better understand the problem, think of the different meanings of the word *architect* in the context of a building architect versus a software architect. If someone types *architect* into a search box, a keyword-based search engine will return a mixed set of documents, with some being about software architects and others about building architects. These kinds of mixed results will frustrate the user who is almost certainly only looking for a specific one of the two meanings. Even if a user searches for *building architect*, typical keyword-based search engines will often transform that query to the boolean query *building AND architect* as independent terms, which may still cause the retrieval of documents that talk about someone who is a *software architect* that is *building* software. Developing smarter search engines to overcome such problems is what we'll discuss throughout the rest of this chapter. With access to the search history of thousands or millions of users, we can discover relationships between search phrases and the most common meaning of each term. Such semantic knowledge can be then be further utilized to better understand the intent of the user.

4.2 Semantic Similarity

Semantic similarity is a measure of the likeness of meaning between two terms [10, 11]. The two major approaches used to compute semantic similarity are through semantic networks (knowledge-based approach) [12], and through computing the relatedness of terms within a large corpus of text (corpus-based approach) [11]. The major techniques classified under corpus-based approaches are Point-wise Mutual Information (PMI) [13] and Latent Semantic Analysis (LSA) [14]. Studies show that PMI typically outperforms LSA on mining synonyms on the web [15].

Another interesting methodology for discovering semantic relationships between words is what Google researchers proposed in [16]. The two novel models proposed by Google are the following:

1. Continuous Bag-of-Words model (CBOW)
2. Continuous Skip-gram model (SG)

These models use large-scale (deep) Neural Networks to learn word vectors. However, the two models are not suitable in our use case due to the a few restrictions. First is the lack of context in our dataset, which is composed of queries that usually contain only 1–3 keywords. The CBOW and SG do not perform well without context, which make our use case challenging. The other limitation is that those models are most suitable for uni-grams or single tokens as opposed to phrases, whereas phrases are most commonly entered by users who conduct searches. For example "*Java Developer*" should be considered as a single phrase when we discover other semantically-related phrases. In our experiment, we discovered high quality semantic relationships using a data set of 1.6 billion search logs entries (made up of keywords used to search for jobs on careerbuilder.com). For this task, we utilized the Probabilistic Graphical Model for Massive Hierarchical Data (PGMHD) [17], which was implemented over the known distributed computing framework Apache Hadoop.

4.3   *Probabilistic Semantic Similarity Scoring Using PGMHD*

The probabilistic-based semantic similarity score is a normalized score between [0,1] that reflects the probability of seeing two terms in the same context. For example, the probabilistic similarity score should reflect that *Java* and *Hadoop* are semantically-related, while *Java* and *Registered Nurse* are not. In order to accomplish this, we utilize the Probabilistic Graphical Model for Massive Hierarchical Data (PGMHD). PGMHD requires collection of the search terms entered by the users to conduct searches, as well as each user's classification. The way to represent this data in order to calculate the probabilistic-based semantic similarity score is to place the classes to which the users belong in the top layer of the model, place the search terms in the lower layer of the model, and then connect them with edges that represent how many users from a given class in the top layer searched for a given term in the lower layer. Table 1 shows the row input data and Fig. 2 shows the representation of that raw data in PGMHD.

Table 1 Input data to PGMHD over hadoop

User1	Java Developer	Java, Java Developer, C#, Software Engineer
User2	Nurse	RN, Registered Nurse, Health Care
User3	.NET Developer	C#, ASP, VB, Software Engineer, SE
User4	Java Developer	Java, JEE, Struts, Software Engineer, SE
User5	Health Care	Health Care Rep, HealthCare

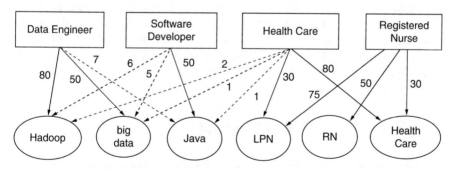

Fig. 2 Using PGMHD to represent job search logs by placing the users' classification at the top layer while the search terms are placed at the lower layer. Each parent node on the top level (job category) stores the number of users classified under that category who conducted searches, while the child nodes (search terms) store the number of times people searched for that term. The edges stores the number of users from the parent node who searched for the term represented by the connected child node

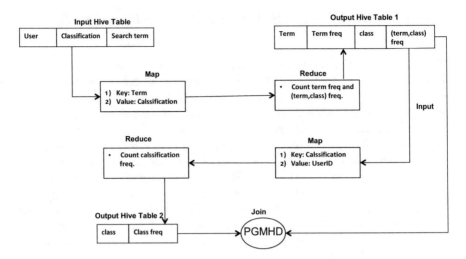

Fig. 3 PGMHD implementation as Map/Reduce using the distributed computing framework Apache Hadoop. The distributed implementation enables PGMHD to represent and process data extracted from 1.6 billion search logs in 45 min

4.4 Distributed PGMHD

In order to process 1.6 billion search log entries (each search log entry contains one or more keywords entered by a user to search for jobs) in reasonable time, we designed a distributed PGMHD using several components of the distributed computing framework Apache Hadoop: HDFS [18], Hadoop Map/Reduce [19], and Hive [20]. The design of distributed PGMHD is shown in Fig. 3. Basically, we use Hive to store the intermediate data while we are building and training the PGMHD. Once it is trained we can then run our inquiries to get an ordered list of the semantically related keywords for any specific term(s).

5 Word Sense Ambiguity Detection

We can utilize the discovered semantically-related terms to improve query understanding. One way to do that is by expanding a submitted query to also include the semantically-related terms, which will help the search engine to retrieve more relevant results since the presence of the query and/or its semantically-related terms in a document will boost that document over the ones which only mentioned the term given in the query. For example, the query *"big data"* can be expanded to *"big data" OR hadoop OR spark OR hive*. As one would expect, the results of the expanded query will typically be more relevant and comprehensive. This technique will not work as intended, however, when dealing with terms that can represent significantly different meanings (ambiguous terms). An *ambiguous term* is a term that refers to more than one meaning depending on the context. For example, the term *java* may refer to the programming language Java, or a type of coffee called java, or an island in Indonesia named Java. Since a user executing a search query is most likely to be searching only for a specific sense of a term, it is important that we can identify and disambiguate between the possible senses. In order to detect those ambiguous terms we again utilize PGMHD, where we calculate a classification score for each term with its parents as potential classes. If the classification score is higher than a specific threshold for more than one parent, we consider that term may be ambiguous one. The idea behind this technique is that each parent class in PGMHD represents a group of users from different classifications, so when a term can be classified with a high confidence score to more than one class, it means it was used widely by users from both classes. Further, if the set of other terms used along with the term varies significantly across multiple classes, this further implies that the term refers to two or more different concepts. Our technique to detect the ambiguous terms is explained below:

Let:

- $C := \{C_1, ..., C_n\}$ be the set of different classes of jobs (Java Developer, Nurse, Accountant, etc);
- $S = \{t_1, ..., t_N\}$ be the set of different search terms entered by users when they conducted searches (N is the number of different terms); and.
- $f(C_j, s)$ be the number times (frequency) a user from class $C_j \in C$ searched for the keyword $s \in S$.

  - To reduce noise, we will only consider the frequencies with at least 100 distinct searches, i.e., $f(c,s) \geq 100$.

$$f(c,s) \geq 100.$$

Then, define

- $O(c)$: the number of times a user from class c searched for a keyword i.e.:

$$O(c): \sum_{s \in S} f(c,s) \quad c \in C;$$

- $T(s)$: the number of times the keyword t_j is searched, i.e.:

$$T(s): \sum_{c \in C} f(c,s) \quad s \in S;$$

- T: the total number of keyword searches, i.e.:

$$T: \sum_{c,s} f(c,s) = \sum_{c \in C} O(c) = \sum_{s \in S} T(s).$$

For every $c \in C$ and $s \in S$, and letting C and S be the random variables representing the class of job and the search term of a single user query, respectively, we can estimate their PMI given by

$$\frac{\mathbb{P}(C=c.S=s)}{\mathbb{P}(C=c)\mathbb{P}(S=s)} = \frac{\mathbb{P}(C=c|S=s)}{\mathbb{P}(C=c)}$$

as follows

$$\mathrm{pmi}(c,s): \log \frac{f(c,s)}{T(s)} \frac{T}{O(c)} \quad c \in C, s \in S.$$

The normalized version [13] of the original PMI estimate is given by

$$\tilde{p}(c,s): \frac{\mathrm{pmi}(c,s)}{-\log \dfrac{f(c,s)}{T}} = \frac{\log T + \log f(c,s) - \log\big[O(c)T(s)\big]}{\log T - \log f(c,s)}$$

$$= -1 + \frac{2\log T - \log O(c) - \log T(s)}{\log T - \log f(c,s)} \in \big[-1,1\big] c \in C, s \in S.$$

This normalized version of the original PMI can then be leveraged to generate an ambiguity score to determine whether or not a term should be considered ambiguous.

5.1   Ambiguity Score

For every search keyword $s \in S$, we define the following *ambiguity score* $A_\alpha(s)$ as

$$A_\alpha(s): \big|\{i: \tilde{p}(c,s) > 0\}\big|,$$

and we say that a search keyword t_j is a *candidate* to be ambiguous if $A_j(\alpha) > 1$. Then, we can define a set of candidate ambiguous terms CA as

$$CA = \{t_j : A_j(\alpha) > 1, j = 1,\dots,N\}.$$

Fig. 4 The proposed
system to resolve word
sense ambiguity using
PGMHD

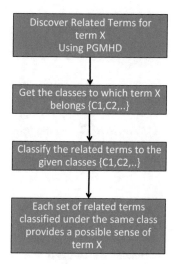

5.2 Resolving Word Sense Ambiguity

After detecting ambiguous terms, the challenge next becomes how to resolve this ambiguity. Resolving ambiguity means defining the possible meanings of an ambiguous term. In our system we leverage the semantically-related terms which we discovered using the previously-discussed semantic discovery module. Each group of those semantically-related terms represents a possible meaning of the original term given the context in which the terms were used when they appeared with that term. For example, the ambiguous term *driver* has semantically-related terms *transportation, truck driver, software, embedded system,* and *CDL*. By classifying these terms using the classes of the users who provided them in the search logs, we end up classifying them into the two groups *"transportation, truck driver, CDL"* and *"software, embedded system"*. It is clear that each of these groups of those semanticallyrelated terms represents a separate possible meaning of driver, with the former group representing the sense of *transportation* and the later instead representing the idea of a *computer device driver*.

Figure 4 shows our methodology to resolve ambiguity. Since we already created a PGMHD for detecting the ambiguous terms, we can utilize the same model to find the semantically-related terms for any given term that falls within the same class. To do so, we calculate the probabilistic-based similarity score between the given term X and a term Y given they both share the same parent class(es) as follows:

Fix a level $i \in \{2,...,m\}$, and let $X, Y \in L_2 \times \cdots \times L_m$ be identically distributed random variables. We define the *probabilistic-based similarity score CO (CoOccurrence)* between two independent siblings $X_{ij}, Y_{ig} \in L_i$ by computing the conditional joint probability

$$CO\left(X_{ij}, Y_{ig}\right) : P\left(X_{ij}, |Y_{ig}, |\mathrm{pa}\left(X_{ij}\right) \cap \mathrm{pa}\left(Y_{ig}\right)\right)$$

as

$$\text{CO}\left(X_{ij},Y_{ig}\right)=\prod_{C'_k \in \text{pa}\left(X_{ij}\right)\cap \text{pa}\left(Y_{ig}\right)}P\left(X_{ij}|C'_k\right)P\left(Y_{ig}|C'_k\right),$$

where $P\left(X_{ij}|C'_k\right)=\dfrac{P\left(C'_k,X_{ij}\right)}{P\left(C'_k\right)}$ for every $\left(X_{ij},C'_k\right)\in L_{i-1}\times L_i$.

Given $out\left(C'_k\right)$ as the total number of occurrences of C'_k and $f\left(C'_k,X_{ij}\right)$ as the frequency of co-occurrence of C'_k with X_{ij}, we can naturally estimate the joint probabilities $P\left(X_{ij},C'_k\right)$ with $\hat{p}\left(X_{ij},C'_k\right)$ defined as

$$\hat{p}\left(X_{ij},C'_k\right):\dfrac{f\left(C'_k,X_{ij}\right)}{out\left(C'_k\right)}$$

Hence, we can estimate the correlation between X_{ij} and Y_{ig} by estimating the probabilistic similarity score $\text{CO}(X_{ij},Y_{ig})$.

Once the list of related terms is generated using PGMHD, we classify them into the classes (since the term is ambiguous, they must belong to more than one class) to which the ambiguous term belongs. This classification phase of the related terms is also implemented using PGMHD as follows:

For a random variable at level $i \in \{2,...m\}$, namely $X_{ij} \in L_i$, where X_{ij} is the jth random variable at level i, we calculate a *classification score* $Cl\left(C'_k|X_{ij}\right)$ for X_{ij} given its primary parent $C'_k \in L_{i-1}$. It is used to estimate the conditional probability $P\left(C'_k|X_{ij}\right)$. The notation C'_k is used to denote a parent, and when it is at level 1, it will represent class C_j as denoted previously. Let

$$f\left(C'_k,X_{ij}\right)=\textit{Frequency of co-occurrence of } C'_k \textit{ and } X_{ij}$$

$$Cl\left(C'_k|X_{ij}\right):\dfrac{f\left(C'_k,X_{ij}\right)}{in\left(X_{ij}\right)}$$

The classification score is the ratio of the co-occurrence frequency of C'_k and X_{ij} divided by the total occurrence of X_{ij}. The total occurrence of X_{ij} is calculated by summing up the frequencies of the co-occurrence of X_{ij} and all its parents.

$$in\left(X_{ij}\right):\sum_{C\in \text{pa}\left(X_{ij}\right)}f\left(C,X_{ij}\right),\qquad \forall X_{ij}\in V,$$

The group of semantically-related terms that get classified under the same parent class will form a possible meaning of the ambiguous term. Using this technique we are not restricted to a limited number of possible meanings: some terms are assigned two possible meanings, some receive three possible meanings, and so on.

6   Semantic Knowledge Graph

In addition to mining query logs to automatically build up semantic knowledge bases, it is also possible to exploit the interrelationship between words and phrases encoded within both the free-text and structured content within a corpus of documents.

Given our focus in this chapter on leveraging big data analytics using large-scale distributed algorithms, our goal is to leverage a system that is able to generate a graph representation of a knowledge domain automatically, merely by ingesting a corpus of data representative of a domain. Once this graph is built, we can then traverse it to surface the interrelationships between each of the the keywords, phrases, extracted entities, and other linguistic variations represented in the corpus. This model is referred to as a Semantic Knowledge Graph [21], and an open source implementation is also publicly available.[1]

Other ontology learning systems typically try to extract specific entities from a corpus and build up a pre-generated graph of relationships between entities. This unfortunately results in a significant loss of information about the nuanced ways in which the meaning of a term or phrase changes depending upon its linguistic context. One of the goals of the Semantic Knowledge Graph approach is to fully preserve all the nuanced semantic interrelationships contained within a textual corpus of documents.

To really understand the significance of this goal, let's consider how the meaning of words can vary depending upon the context in which they are found. The words *architect* and *engineer* are well known, but when found inside phases such as *software architect* or *electrical engineer*, they take on a much more limited interpretation. Similarly, the word *driver* can take on numerous different meanings, such as when found near terms relating to computers (a *hardware driver*), a golf game (a kind of golf club), a business analysis ("a key *driver* of costs"), or in contexts related to transportation (*truck driver* or *delivery driver*). Even when focused on transporting goods, the word *driver* will have a nuanced difference in meaning in the context of a night club (a taxi to safely transport someone home), a hospital (some kind of medical transport), or on a race track (a competitor trying to outrun other vehicles). While people typically think that most words have a limited number of meanings, it is more accurate to consider words and phrases as having a different meaning in every possible context in which they appear (even if the difference is nuanced). While the intended meaning of words and phrases across different contexts will all share strong similarities, the Semantic Knowledge Graph is able to model those similarities while also preserving each of the context-dependent nuances in meaning. By surfacing these nuanced meanings of words and phrases during node traversals, the Semantic Knowledge Graph is thus able to better represent the entire underlying knowledge domain in a compact and highly context-aware representation.

---

[1] https://github.com/careerbuilder/semantic-knowledge-graph

6.1 Model Structure

Given an undirected graph $G = (V,E)$ with V and $E \subset V \times V$ denoting the sets of nodes and edges, respectively, we establish the following definitions:

- $D = \{d_1, d_2, ..., d_m\}$ is the set of documents that represents a corpus that we will utilize to identify and score semantic relationships within the Semantic Knowledge Graph.
- $X = \{x_1, x_2, ..., x_k\}$ is the set of all items which are stored in D. These items may be terms, phrases, or even any arbitrary linguistic representations that can be found within D.
- $d_i = \{x | x \in X\}$ where each document $d \in D$ is a set of items.
- $T = \{t_1, t_2, ..., t_n\}$ where t_i is a tag that identifies an entity type for an item. Examples of tags may include keyword, location, school, company, person, etc.

Given these definitions, the set of nodes V in the graph is defined as $V = \{v_1, v_2, ..., v_n\}$ where v_i represents an item $x_i \in X$ tagged with tag $t_j \in T$, while $D_{vi} = \{d | x_i \in d, d \in D\}$ is the set of documents containing item x_i with its corresponding tag t_j. We then define e_{ij} as the edge between (v_i, v_j) by a function $f(e_{ij}) = \{d \in D_{vi} \cap D_{vj}\}$ that represents each edge with the set of documents containing both item x_i and item x_j, each with their corresponding tags. Finally, we define a function $g(e_{ij}, v_k) = \{d: d \in f(e_{ij}) \cap D_{vk}\}$ that stores the common set of documents between $f(e_{ij})$ and D_k on each edge e_{jk}.

6.2 Materialization of Nodes and Edges

The SKG model differs from most traditional graph structures by leveraging a layer of indirection between any two nodes and the edge that connects them. Specifically, instead of two nodes v_i and v_j being directly connected to each other through an explicit edge e_{ij}, nodes are instead connected through documents, such that the edge e_{ij} between node v_i and v_j is said to *materialize* any time $|f(e_{ij})| > 0$.

In order to traverse from a source node v_i to another node v_j, our system thus requires a lookup index (an inverted index) that maps node v_i to an underlying set of documents, as well as different lookup index (a forward index) that is able to map those those documents to any other node v_j to which those documents are also linked. This combination of inverted index and forward index allows all terms or combinations of terms to be modeled as nodes in the graph, enabling the traversal between any two nodes through the set of shared documents between them, as shown in Fig. 5.

Since edges are based upon a set intersection of the documents both nodes are linked to, this means that an edge can also be generated on the fly between *any* arbitrary combination of other nodes. We refer to this dynamic generation of edges as *materialization* of edges. Further, because both nodes and edges are based entirely on set intersections of documents, this means it is also possible to dynamically materialize new nodes based upon arbitrary combinations of other nodes, as shown in Fig. 6.

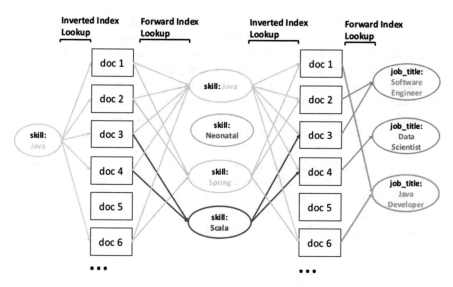

Fig. 5 Materialization of edges using shared documents. Edges exist between documents which share terms. The edge weights are calculated on the fly using a function that leverages the statistical distribution of documents shared between the nodes

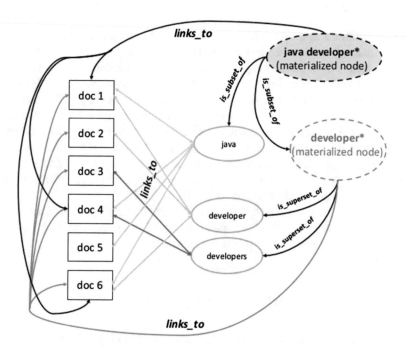

Fig. 6 Materializing new nodes dynamically. New nodes can be formed dynamically from any arbitrary combination of other nodes, words, phrases, or any other linguistic representation

Since both nodes and edges can be materialized on the fly, this not only enables us to generate nodes representing arbitrarily-complex combinations of existing terms, but also to decompose arbitrarily-complex entities and relationships into their constituent parts. For example, we can store just the nodes *software* and *engineer* in the inverted index and forward index (along with positional information about where they appear in each document), knowing that we can easily reconstruct the longer phrase *"software engineer"* later as a materialized node. We can even reconstruct arbitrarily-complex nodes such as *"software engineer* in in the location of *New York* that also have the skills of *Java* and *Python* and either the words *contract* or *contractor* or *work to hire* or the word *negotiable* within three words of *pay* or *salary"*. The Semantic Knowledge Graph, therefore, provides both a lossless and yet highlycompressed representation of every possible linguistic variation found within the original corpus, as well as every potential edge that could connect all possible materialized nodes with other nodes.

6.3   Discovering Semantic Relationships

One of the key capabilities of the semantic knowledge graph is its ability to uncover hidden relationships between nodes. In order to discover a relationship between a node with a specific tag (field name) t_k to another item x_i with a specific tag t_j, we first query the inverted index item x_i and assign its document set to node v_i corresponding with the document set D_{vi}. To then find the candidate nodes to which we should traverse, we then search the forward index for tag t_k, and we reference this set of matching documents as $D_{tk} = \{d | x \in d, x: t_k\}$. We then define $V_{vi \cdot tk} = \{v_j | x_j \in d, d \in D_{tk} \cap D_{vi}\}$ with v_j being the node that stores item x_j, and we further define $V_{vi \cdot tk}$ as the set of nodes storing items with an edge to x_i of type t_k (see Fig. 7). We then apply $\forall v_j \in V_{vi \cdot tk}, relatedness_{(vi \cdot vj)}$ in order to score the semantic relationship between v_i and v_j. This *relatedness* score, which will be described in the next subsection, enables us to rank each of the edges between nodes in order to pick the top m most related nodes. We can also define a threshold t in order to only accept relationships with *relatedness*$(v_i, v_j) > t$. This above operation can occur recursively in order to traverse into multiple levels of relationships, as shown in Fig. 8.

The weights are calculated based upon the entire traversed path here, though it is possible to alternatively calculate weights not conditioned upon the path and using only each separate pair of directly connected nodes.

6.4   Scoring Semantic Relationships

One of the most powerful features of the Semantic Knowledge Graph (SKG) is its ability to score the edges between nodes in the graph based upon the strength of the semantic similarity between the entities represented by those nodes. If we don't know how related the phrase *physician's assistant* is to the keyword *doctor* or even

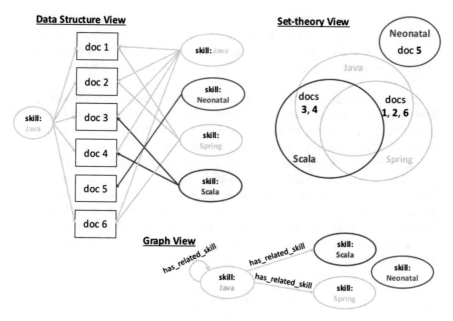

Fig. 7 Three representations of a traversal. The Data Structure View represents the underlying links from term to document to term in our underlying data structures, the Set Theory view shows the relationships between each term once the underlying links have been resolved, and the Graph View shows the abstract graph representation in the semantics exposed when interacting with the SKG

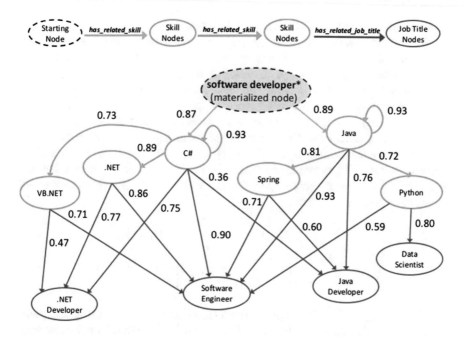

Fig. 8 Graph traversal. This example traverses from a materialized node (*software developer**), through all *has-related-skill* edges, then from each node at that level again through their *has related skill* edges, and finally from those nodes to each of their *has related job title* edges

the phrase *truck driver*, we can leverage the SKG to score the strength of the semantic relationship between all of those terms. To calculate the semantic similarity score between items x_i and x_j, we materialize a source node v_i (representing the document set containing x_i) and destination node v_j (representing the document set containing x_j).

The simplest example of scoring semantic relationships is when comparing two directly connected nodes, which we'll call v_i and v_j. To do this, we first query the inverted index for item x_i, which is tagged with t_j, and this query returns back D_{vi}. We then perform a similar query for x_j, which is tagged with t_k, which returns back D_{vj}. An edge e_{ij} exists between v_i and v_j when $f(e_{ij})$ 6 = φ. We refer to D_{vi} as our *foreground* document set D_{FG} and correspondingly call $D_{BG} \subseteq D$ our *background* document set. Our scoring technique relies upon the hypothesis that x_i is more semantically-related to x_j when the relative frequency of x_j occurring in the foreground document set D_{FG} is greater the the relative frequency of x_j occurring in the background document set D_{BG}. We leverage the z score as our similarity measure for this hypothesis:

$$z\left(v_i,v_j\right) = \frac{y - n * p}{\sqrt{n * p\left(1 - p\right)}}$$

Where $n = |D_{FG}|$ is the size of the foreground document set, $y = |f(e_{ij})|$ is the count of documents that contain both x_i and x_j, and $p = \dfrac{\left|D_{v_j}\right|}{D_{BG}}$ is the probability of seeing term x_j with tag t_k within the background document set.

We often may want to traverse the graph more than one level of depth to score the relationships between more than two nodes, however. If we chose to traverse from the entity *java* to *developer* to *architect*, for example, the weight of the edge between *developer* and *architect* would make more contextual sense if it were also conditioned upon the previous path traversed from *java* to *developer*. Otherwise, the nuanced difference in meaning of the word *architect* in this context is lost in the edge scoring. The Semantic Knowledge Graph enables us to retain this context from any previous n nodes along a path $P = v_1, v_2, ..., v_n$, with each node storing an item x_i having a tag t_j. To calculate the same $z(v_i, v_j)$ between any two nodes, but also conditioning the edge's score upon the full path P, the following changes are required to the scoring function:

$$D_{FG} = \begin{cases} f\left(e_{ij}\right) & \text{if } n = 3 \\ \left\{ \bigcap\limits_{i=1, j=i+1, k=j+1}^{n-3} g\left(e_{ij}, D_{v_k}\right) \right\} & \text{if } n > 3 \end{cases}$$

where $y = \left|D_{FG} \cap D_{v_n}\right|$. We apply normalization on the z score using a sigmoid function such that the scores fall within the range $[-1, 1]$. We refer to this normalized score between nodes as their *relatedness score*, where 1 indicates a completely positive relationship (very likely to appear together), where 0 means no relationship (unrelated and just as likely as any random node to appear together), and where −1 means a completely negative relationship (highly unlikely to appear together).

It is important to note that since the edge weights are calculated at traversal time (edges are materialized), that it is possibly to easily substitute a different scoring function when appropriate. A simpler, but typically less meaningful, alternate scoring function would be the total count of overlapping documents, which is what most graph databases tend use for edge scoring. Plugging in more complex scoring functions leveraging the statistics available in the inverted index and forward index is also possible.

6.5 Scaling Characteristics

The Semantic Knowledge Graph, being built on top of an inverted index and forward index, fundamentally shares the same scaling characteristics of the underlying distributed search engine.

As described in Sect. 1, both the inverted index and forward index data structures scale well horizontally to trillions of documents sharded across multiple servers. While there will be heavy overlap between the terms in every shard of the inverted index and forward index, the number of terms conveniently grows logarithmically, since each additional document is less likely than the last to add new terms to the index that were never seen in a previous document. The documents, conversely, are always partitioned across servers, such that all operations can occur in parallel against only the subset of documents on each shard. Once these distributed operations are completed, then only one final aggregation of the top results from each shard is necessary to return a final result.

For multi-relationship graph traversals (i.e. traverse from *skills* to *job titles* and then also to *industries*), it is necessary for an additional aggregation to occur for each nested level of traversal. This *refinement* process is to ensure that no nodes (terms) were missed due to not being returned from one or more shards. For example, if we run a graph traversal across two shards and shard 1 returns the nodes a, b, c, but shard 2 returns nodes a, c, d, then it is necessary to send another refinement request to shard 1 to return its statistics for the previously missing node d and one request to shard 2 to return its statistics for the previously missing node b.

This refinement cost scales linearly with the number of nested levels requested, and it should be uncommon to have many nested traversal levels for most common use cases. Given these scaling characteristics, the Semantic Knowledge Graph can be easily built and run at massive scale to enable distributed graph traversals across a massive semantic knowledge base.

7 Real World Applications

We implemented the techniques described throughout this chapter within the context of a career search website. Specifically, they were implemented as components of a semantic search system for CareerBuilder, one of the largest online job boards in the world. The system leveraged the described query log mining techniques (as described in 4.2) to build up a language-agnostic and domain specific taxonomy that was able to model and disambiguate words (as described in 5) and related terms, as well as the Semantic Knowledge Graph, which could also discover and score the strength of named relationships between terms. By combining both a user-input-based approach (mining query logs) and a content-based approach (as described in 6), we were able to improve the quality of the output of both systems. For example, we were able to use the Semantic Knowledge Graph to score the terms and coterms found from mining the query logs, enabling us to reduce the noise in the coterms lists with 95% accuracy [21]. While the usefulness of the related coterms was higher in the list mined from query logs (because the logs directly model the language used by users of the system to express their intent), the Semantic Knowledge Graph was able to fill in holes in the learned taxonomy for terms or coterms which were not adequately represented within the query logs.

For our production system, we ended up indexing all discovered terms into a scalable, naive entity extractor called the Solr Text Tagger.[2] The Solr Text Tagger leverages Apache Solr to build an inverted index compressed into a specialized data structure called a Finite State Transducer (FST). This data structure enables us to index millions of potential entities and subsequently pass incoming queries and documents in to perform entity extraction in milliseconds across reasonably large documents.

The extracted entities can then be passed to the Semantic Knowledge Graph in order to score their similarity with the topic of the document. This allows us to take, for example, a 10,000 word document and summarize it using the top ten phrases which are most relevant to that document. It is then possible to run a weighted search for those top keywords to find a relevant set of related documents (which provides a highly accurate content-based recommendation algorithm), or to alternatively traverse from those top ten phrases to a list of phrases most relevant to them, but potentially missing from the actual document. In this way, we can search on the concepts people are looking for, without relying on the exact words they have used within their documents.

The same process of entity extraction, ranking, and concept expansion that we described for documents also works well for interpreting and expanding queries in order to provide a powerful semantic search experience. This system, in production, was able to boost the NDCG scores (which is common metric used to measure relevancy of a search engine) of search results from 59–76%, representing a very significant improvement in the relevancy of the search engine [21].

---

[2] https://github.com/OpenSextant/SolrTextTagger

8   Conclusion

We have discussed many techniques and tools available for building and utilizing semantic knowledge bases. These techniques include the mining of massive volumes of query logs leveraging a Probabilistic Graphical Model for Massive Hierarchical Data (PGMHD) across a Hadoop cluster to find interesting terms and phrases along with semantically-related terms and phrases which can be used for concept expansion [22]. We also described a method for detection and disambiguation of multiple senses of those discovered terms and phrases found within the query logs [23]. We further covered a model called a Semantic Knowledge Graph, which leverages the relationships inherent between words and phrases within a corpus of documents to automatically generate a relationship graph between those phrases. This graph can be traversed to further discover and score the strength of relationships between any entities contained within it based purely upon the content within the documents in a search engine.

These components by themselves are useful tools, but when combined together, they can form a powerful "intent engine" which is able to index content into a search engine, and then leverage the auto-generated semantic knowledge bases to parse and interpret incoming queries (to match documents) or documents (to match other documents). We successfully applied these techniques at one of the largest job boards in the world and were ultimately able to boost the relevancy of the search engine (as measured by NDCG scores) from 59–76%. Such a significant improvement in search results relevancy is a testament to the gains which can be achieved through utilizing distributed big data analytics to automate the creation of semantic knowledge bases and applying them to increase the relevancy of an information retrieval system.

References

1. R. Navigli and P. Velardi, "Learning domain ontologies from document warehouses and dedicated web sites," *Computational Linguistics*, vol. 30, no. 2, 2004.
2. T. Grainger and T. Potter, *Solr in Action*. Manning Publications Co, 2014.
3. J. Bobadilla, F. Ortega, A. Hernando, and A. Gutierrez,´ "Recommender systems survey," *Knowledge-Based Systems*, vol. 46, pp. 109–132, 2013.
4. J. Lu, D. Wu, M. Mao, W. Wang, and G. Zhang, "Recommender system application developments: a survey," *Decision Support Systems*, vol. 74, pp. 12–32, 2015.
5. C. C. Aggarwal, "Content-based recommender systems," in *Recommender Systems*, pp. 139–166, Springer, 2016.
6. M. J. Pazzani and D. Billsus, "Content-based recommendation systems," in *The adaptive web*, pp. 325–341, Springer, 2007.
7. X. Su and T. M. Khoshgoftaar, "A survey of collaborative filtering techniques," *Advances in artificial intelligence*, vol. 2009, p. 4, 2009.
8. R. Burke, "Hybrid recommender systems: Survey and experiments," *User modeling and user-adapted interaction*, vol. 12, no. 4, pp. 331–370, 2002.

9. M. de Gemmis, P. Lops, C. Musto, F. Narducci, and G. Semeraro, "Semantics-aware content-based recommender systems," in *Recommender Systems Handbook*, pp. 119–159, Springer, 2015.

10. S. Harispe, S. Ranwez, S. Janaqi, and J. Montmain, "Semantic measures for the comparison of units of language, concepts or entities from text and knowledge base analysis," *arXiv preprint arXiv:1310.1285*, 2013.

11. R. Mihalcea, C. Corley, and C. Strapparava, "Corpus-based and knowledge-based measures of text semantic similarity," in *AAAI*, vol. 6, pp. 775–780, 2006.

12. A. Budanitsky and G. Hirst, "Semantic distance in wordnet: An experimental, applicationoriented evaluation of five measures," in *Workshop on WordNet and Other Lexical Resources*, vol. 2, 2001.

13. G. Bouma, "Normalized (pointwise) mutual information in collocation extraction," in *Proceedings of the Biennial GSCL Conference*, pp. 31–40, 2009.

14. S. T. Dumais, "Latent semantic analysis," *Annual review of information science and technology*, vol. 38, no. 1, pp. 188–230, 2004.

15. P. D. Turney, "Mining the web for synonyms: PMI-IR versus lsa on toefl," in *Proceedings of the 12th European Conference on Machine Learning*, EMCL '01, (London, UK, UK), pp. 491–502, Springer-Verlag, 2001.

16. T. Mikolov, K. Chen, G. Corrado, and J. Dean, "Efficient estimation of word representations in vector space," *arXiv preprint arXiv:1301.3781*, 2013.

17. K. AlJadda, M. Korayem, C. Ortiz, T. Grainger, J. A. Miller, and W. S. York, "Pgmhd: A scalable probabilistic graphical model for massive hierarchical data problems," in *Big Data (Big Data), 2014 IEEE International Conference on*, pp. 55–60, IEEE, 2014.

18. K. Shvachko, H. Kuang, S. Radia, and R. Chansler, "The hadoop distributed file system," in *Mass Storage Systems and Technologies (MSST), 2010 IEEE 26th Symposium on*, pp. 1–10, IEEE, 2010.

19. J. Dean and S. Ghemawat, "Mapreduce: simplified data processing on large clusters," *Communications of the ACM*, vol. 51, no. 1, pp. 107–113, 2008.

20. A. Thusoo, J. S. Sarma, N. Jain, Z. Shao, P. Chakka, S. Anthony, H. Liu, P. Wyckoff, and R. Murthy, "Hive: a warehousing solution over a map-reduce framework," *Proceedings of the VLDB Endowment*, vol. 2, no. 2, pp. 1626–1629, 2009.

21. T. Grainger, K. AlJadda, M. Korayem, and A. Smith, "The semantic knowledge graph: A compact, auto-generated model for real-time traversal and ranking of any relationship within a domain," in *IEEE 3rd International Conference on Data Science and Advanced Analytics*, IEEE, 2016.

22. K. AlJadda, M. Korayem, T. Grainger, and C. Russell, "Crowdsourced query augmentation through semantic discovery of domain-specific jargon," in *IEEE International Conference on Big Data (Big Data 2014)*, pp. 808–815, IEEE, 2014.

23. M. Korayem, C. Ortiz, K. AlJadda, and T. Grainger, "Query sense disambiguation leveraging large scale user behavioral data," in *IEEE International Conference on Big Data (Big Data 2015)*, pp. 1230–1237, IEEE, 2015.

Printed in the United States
By Bookmasters